Electricity and New Energy

AC/DC Motors and Generators

Student Manual

584108

Order no.: 584108
Revision level: 12/2014

By the staff of Festo Didactic

© Festo Didactic Ltée/Ltd, Quebec, Canada 1996, 2010
Internet: www.festo-didactic.com
e-mail: did@de.festo.com

Printed in Canada
All rights reserved
ISBN 978-2-89640-422-3 (Printed version)
ISBN 978-2-89747-161-3 (CD-ROM)
Legal Deposit – Bibliothèque et Archives nationales du Québec, 2010
Legal Deposit – Library and Archives Canada, 2010

The purchaser shall receive a single right of use which is non-exclusive, non-time-limited and limited geographically to use at the purchaser's site/location as follows.

The purchaser shall be entitled to use the work to train his/her staff at the purchaser's site/location and shall also be entitled to use parts of the copyright material as the basis for the production of his/her own training documentation for the training of his/her staff at the purchaser's site/location with acknowledgement of source and to make copies for this purpose. In the case of schools/technical colleges, training centers, and universities, the right of use shall also include use by school and college students and trainees at the purchaser's site/location for teaching purposes.

The right of use shall in all cases exclude the right to publish the copyright material or to make this available for use on intranet, Internet and LMS platforms and databases such as Moodle, which allow access by a wide variety of users, including those outside of the purchaser's site/location.

Entitlement to other rights relating to reproductions, copies, adaptations, translations, microfilming and transfer to and storage and processing in electronic systems, no matter whether in whole or in part, shall require the prior consent of Festo Didactic GmbH & Co. KG.

Information in this document is subject to change without notice and does not represent a commitment on the part of Festo Didactic. The Festo materials described in this document are furnished under a license agreement or a nondisclosure agreement.

Festo Didactic recognizes product names as trademarks or registered trademarks of their respective holders.

All other trademarks are the property of their respective owners. Other trademarks and trade names may be used in this document to refer to either the entity claiming the marks and names or their products. Festo Didactic disclaims any proprietary interest in trademarks and trade names other than its own.

Safety and Common Symbols

The following safety and common symbols may be used in this manual and on the equipment:

Symbol	Description
⚠ DANGER	**DANGER** indicates a hazard with a high level of risk which, if not avoided, will result in death or serious injury.
⚠ WARNING	**WARNING** indicates a hazard with a medium level of risk which, if not avoided, could result in death or serious injury.
⚠ CAUTION	**CAUTION** indicates a hazard with a low level of risk which, if not avoided, could result in minor or moderate injury.
CAUTION	**CAUTION** used without the *Caution, risk of danger* sign ⚠, indicates a hazard with a potentially hazardous situation which, if not avoided, may result in property damage.
⚡	Caution, risk of electric shock
🔥	Caution, hot surface
⚠	Caution, risk of danger
🏋	Caution, lifting hazard
✋	Caution, hand entanglement hazard
📡	Notice, non-ionizing radiation
⎓	Direct current
∿	Alternating current
⌁	Both direct and alternating current
3∿	Three-phase alternating current
⏚	Earth (ground) terminal

Safety and Common Symbols

Symbol	Description
⏚	Protective conductor terminal
⏚	Frame or chassis terminal
⏚	Equipotentiality
│	On (supply)
○	Off (supply)
▣	Equipment protected throughout by double insulation or reinforced insulation
⎕	In position of a bi-stable push control
⎕	Out position of a bi-stable push control

Table of Contents

Preface XI

About This Manual XIII

Unit 1   **Fundamentals for Rotating Machines** 1

An introduction to rotating machines. Work, speed, torque, and power. Operation of the prime mover/dynamometer module. Motor losses and efficiency.

Ex. 1-1   **Prime Mover and Brake Operation (Model 8960-2)** 11

Familiarization with the Four-Quadrant Dynamometer/Power Supply operating in the prime mover mode: measurement of the speed and the opposition torque produced by the driven machine. Familiarization with the Four-Quadrant Dynamometer/Power Supply operating in the brake mode: measurement of the speed and output torque of a drive motor with a brake and a dynamometer.

Ex. 1-2   **Prime Mover and Brake Operation (Model 8960-1)** 29

Familiarization with the Prime Mover/Dynamometer operating in the prime mover mode: measurement of the speed and the opposition torque produced by the driven machine. Familiarization with the Prime Mover/Dynamometer operating in the brake mode: measurement of the speed and output torque of a drive motor with a brake and a dynamometer.

Ex. 1-3   **Motor Power, Losses, and Efficiency** 45

Determining the mechanical output power of a motor from the speed and torque. Mechanical and electrical losses in motors. Determining the efficiency of a motor.

Unit 2   **DC Motors and Generators** 61

The operating principles of direct current (DC) motors and generators. The different types of dc motors and generators and their particularities.

Ex. 2-1   **The Separately-Excited DC Motor** 67

Operation of a separately-excited dc motor. Simplified equivalent circuit of a DC motor. Relationship between the no-load speed and the armature voltage. Relationship between the motor torque and the armature current. Armature resistance. Speed-torque characteristic.

Table of Contents

Ex. 2-2 Separately-Excited, Series, Shunt, and Compound DC Motors 81

Effect of the field current on the speed-voltage and torque-current characteristics of a separately-excited dc motor. Description and operation of the series, shunt, and compound dc motors. Comparing the speed-torque characteristics of the separately-excited, series, shunt, and compound dc motors.

Ex. 2-3 Separately-Excited, Shunt, and Compound DC Generators 99

Operation and characteristics of a separately-excited dc generator. Effect of the field current on the characteristics of a separately-excited DC generator. Simplified equivalent circuit of a DC generator. Operation and characteristics of self-excited DC generators. Comparing the voltage-current characteristics of the separately-excited, shunt, cumulative-compound, and differential-compound dc generators.

Unit 3 Special Characteristics of DC Motors 121

The behaviour of dc machines when the armature and field currents exceed the nominal values. Operation of the universal motor.

Ex. 3-1 Armature Reaction and Saturation Effect 123

Armature reaction. Effect of the armature reaction on the characteristics of DC machines. Armature inductance. Use of permanent-magnets to reduce armature reaction. Saturation. Effect of the saturation on the characteristics of DC machines.

Ex. 3-2 The Universal Motor 139

Direction of rotation versus the polarities of the armature and field currents. DC and AC operation of a universal motor. Improving AC operation by adding a compensating winding that reduces the armature inductance.

Unit 4 AC Induction Motors 157

The principles of electromagnetic induction. Rotating magnetic field and synchronous speed. Demonstrating the operation and characteristics of AC induction motors.

Table of Contents

Ex. 4-1 The Three-Phase Squirrel-Cage Induction Motor 159

Creating a rotating magnetic field in a three-phase squirrel-cage induction motor. Synchronous speed. Description and operation of the three-phase squirrel-cage induction motor. Torque versus speed characteristic. Reactive power required for creating the rotating magnetic field.

Ex. 4-2 Eddy-Current Brake and Asynchronous Generator 171

Description and operation of the eddy-current brake. Operating a three-phase squirrel-cage induction motor as an asynchronous generator. Demonstrating that an asynchronous generator can supply active power to the AC power network. Demonstrating that asynchronous generator operation requires reactive power.

Ex. 4-3 Effect of Voltage on the Characteristics of Induction Motors 185

Saturation in induction motors. Nominal voltage of a squirrel-cage induction motor. Demonstrating the effect of the motor voltage on the torque versus speed characteristic of a squirrel-cage induction motor.

Ex. 4-4 Single-Phase Induction Motors 197

Description and operation of a simplified single-phase squirrel-cage induction motor. Torque-speed characteristic of the simplified single-phase induction motor. Adding an auxiliary winding (with or without a capacitor) to improve the starting torque of the simplified single-phase induction motor.

Unit 5 Synchronous Motors 213

Description and operation of the three-phase synchronous motor. Starting a synchronous motor. Speed of rotation versus the AC power source frequency.

Ex. 5-1 The Three-Phase Synchronous Motor 215

Interesting features of the three-phase synchronous motor. Effect of the field current on the reactive power exchanged between a three-phase synchronous motor and the ac power network. Using a synchronous motor running without load as a synchronous condenser.

Table of Contents

Ex. 5-2 Synchronous Motor Pull-Out Torque 225

Effect of the field current on the pull-out torque of a three-phase synchronous motor.

Unit 6   **Three-Phase Synchronous Generators (Alternators) 235**

Principle of operation of synchronous generators. Description and operation of the three-phase synchronous generator. Three-phase synchronous generator characteristics. Frequency and voltage regulation. Generator synchronization.

Ex. 6-1 Synchronous Generator No-Load Operation 237

Relationship between the speed of rotation and the voltage and frequency of a synchronous generator operating without load. Relationship between the field current and the voltage produced by a synchronous generator operating without load. Saturation in synchronous generators.

Ex. 6-2 Voltage Regulation Characteristics 247

Simplified equivalent circuit of a synchronous generator. Voltage regulation characteristics of a synchronous generator for resistive, inductive, and capacitive loads.

Ex. 6-3 Frequency and Voltage Regulation 257

Effect of resistive, inductive, and capacitive loads on the output voltage and frequency of a synchronous generator. Adjusting the speed and field current of a synchronous generator to regulate its frequency and voltage when the load fluctuates.

Ex. 6-4 Generator Synchronization 265

Conditions to be respected before connecting a synchronous generator to the AC power network or another generator. Adjusting the torque applied to the shaft of a synchronous generator to set the amount of active power it delivers. Adjusting the field current of a synchronous generator to set the power factor to unity.

Appendix A   Circuit Diagram Symbols 277

Appendix B   Impedance Table for the Load Modules 283

Appendix C   Equipment Utilization Chart 287

Table of Contents

Appendix D New Terms and Words **289**

Index of New Terms 291
Bibliography 293

Preface

Computer-based teaching technologies are becoming more and more widespread in the field of education, and the Data Acquisition and Control for Electromechanical Systems (LVDAC-EMS), the Data Acquisition and Management for Electromechanical Systems (LVDAM-EMS), and the Simulation Software for Electromechanical Systems (LVSIM®-EMS) are witness to this new approach.

The LVDAC-EMS (or LVDAM-EMS) system is a complete set of measuring instruments that runs on a Pentium-type personal computer under the Microsoft® Windows® operating environment. Computer-based instruments (voltmeters, ammeters, power meters, an oscilloscope, a phasor analyzer, and an harmonic analyzer) provide instructors the opportunity to clearly demonstrate concepts related to electric power technology that, until now, could only be presented using traditional textbook methods and static drawings.

The LVDAC-EMS (or LVDAM-EMS) system uses a customized data acquisition module to interconnect modules of the Electromechanical System with the personal computer. Dedicated software routes the measured values from the data acquisition module to the computer-based instruments that provide all the standard measurements associated with voltage, current, power, and other electrical parameters. However, the system does much more: it provides built-in capabilities for waveform observation and phasor analysis, data storage and graphical representation, as well as programmable meter functions, thereby allowing unimagined possibilities for presenting courseware material.

LVSIM®-EMS is a software that faithfully simulates the Electromechanical System (EMS). Like the LVDAC-EMS (or LVDAM-EMS) system, LVSIM®-EMS runs on a PC-type computer under the Microsoft® Windows® operating environment.

LVSIM®-EMS recreates a three-dimensional classroom laboratory on a computer screen. Using the mouse, students can install an EMS training system in this virtual laboratory, make equipment setups, and perform exercises in the same way as if actual EMS equipment were used. The EMS equipment that can be installed in the virtual laboratory faithfully reproduces the actual EMS equipment included in the Computer-Assisted 0.2-kW Electromechanical Training System (Model 8006) in every detail. As for the actual EMS system, the operation and behaviour of the circuits simulated with LVSIM®-EMS can be observed by performing voltage, current, speed, and torque measurements, using the same computer-based instruments as for the LVDAC-EMS (or LVDAM-EMS) system.

The existing EMS courseware has been completely revised and adapted for the LVDAC-EMS (or LVDAM-EMS) system as well as LVSIM®-EMS, and the new series is titled Electrical Power Technology Using Data Acquisition. Exercises have been grouped in two separate manuals: manual 1, titled *Power Circuits and Transformers*, and manual 2, titled *AC/DC Motors and Generators*.

Each exercise approaches the subject matter from a practical point of view, and uses a hands-on approach to the study of electrical power technology. Students are guided through step-by-step exercise procedures that confirm concepts and theory presented in the exercise discussion. A conclusion and set of review questions complete each exercise, and a 10-question unit test helps evaluate knowledge gained in the courseware unit.

Preface

Do you have suggestions or criticism regarding this manual?

If so, send us an e-mail at did@de.festo.com.

The authors and Festo Didactic look forward to your comments.

About This Manual

The 18 exercises in this manual, *AC/DC Motors and Generators*, provide a foundation for further study of rotating machines.

This manual is divided into six units:

- Unit 1 provides a basic review of concepts and theory of rotating machines, torque, and speed, as well as highlighting specific details relating to power, losses, and efficiency of electric motors. It also describes the operation of the prime mover and brake used throughout the hands-on exercises. One exercise in Unit 1 focuses on the implementation of the prime mover and brake using the Four-Quadrant Dynamometer/Power Supply, Model 8960-2. Another exercise in Unit 1 focuses on the implementation of the prime mover and brake using the Prime Mover/Dynamometer, Model 8960-1. The student performs either one of these two exercises, depending on whether he or she is using Model 8960-2 or 8960-1.

- Units 2 and 3 deal with the basic operation and characteristics of direct current motors and generators, and explore some of the particularities of dc machines.

- Units 4, 5, and 6 define and explain the concepts related to alternating current motors and generators. The operation of induction motors as well as that of synchronous motors and generators (alternators) are covered.

The hands-on exercises in this manual can be performed using either the Electromechanical System (EMS system) or the Electromechanical System using Virtual Laboratory Equipment (LVSIM®-EMS). When using the EMS system, you should turn on the computer and start Windows® before each exercise. On the other hand, when using LVSIM®-EMS, you should turn on the computer, start Windows®, and start LVSIM®-EMS before each exercise.

The hands-on exercises guide students through circuit setup and operation, and uses many of the measurement and observation capabilities of the virtual instrumentation system. Much detailed information about rotating machine parameters (voltages and currents, torque and speed, output power, and efficiency, etc.) can be visualized with the computer-based instruments, and students are encouraged to fully explore system capabilities.

Various symbols are used in many of the circuit diagrams given in the exercises. Each symbol is a functional representation of a device used in Electrical Power Technology. The use of these symbols greatly simplifies the circuit diagrams by reducing the number of interconnections shown, and makes it easier to understand circuit operation. Appendix A lists the symbols used, the name of the device which each symbol represents, and a diagram showing the equipment and connections required to obtain the device.

The exercises in this manual can be carried out with ac network voltages of 120 V, 220 V, and 240 V. The component values used in the different circuits often depend on the ac line voltage. For this reason, components in the circuit diagrams are identified where necessary with letters and subscripts. A table accompanying the circuit diagram indicates the component value required for each ac network voltage (120 V, 220 V, and 240 V).

About This Manual

Appendix A consists of diagrams showing the equipment and the connections required to obtain the devices used in the exercises.

Appendix B provides a table giving the usual impedance values that can be obtained with each of the 120-V, 220-V, and 240-V versions of the EMS load modules.

Appendix C provides a chart outlining the exact equipment required for each exercise.

Appendix D is a glossary of the new terms and words used in this manual.

Safety considerations

Safety symbols that may be used in this manual and on the equipment are listed in the Safety Symbols table at the beginning of the manual.

Safety procedures related to the tasks that you will be asked to perform are indicated in each exercise.

Make sure that you are wearing appropriate protective equipment when performing the tasks. You should never perform a task if you have any reason to think that a manipulation could be dangerous for you or your teammates.

Unit 1

Fundamentals for Rotating Machines

UNIT OBJECTIVE

When you have completed this unit, you will be able to explain motor and generator operation using basic concepts of magnetism. You will also be able to demonstrate torque, speed, and mechanical power measurements with the prime mover/dynamometer module.

DISCUSSION OF FUNDAMENTALS

Everyone is familiar with some type of **electric motor** or another, whether it is the tiny **dc motor** in battery-operated toys, the dc starter motor in automobiles, or the **ac motor** in washing machines and clothes dryers. Electric motors are also used in fans, electric drills, pumps and many other familiar devices. But how and why do these motors work, and why do they turn? The answer is surprisingly simple; it is because of the interaction between two magnetic fields.

If you take two magnets and fix one of them on a shaft so it can rotate, and then move the second magnet in a circle around the first, the **rotor** magnet will be pulled along because of the **magnetic force** of attraction between the two, as shown in Figure 1-1a. As a result, the rotor magnet will rotate in synchronization with the pulling magnet.

This simple image of the interaction between two magnets is shown in a more realistic way with Figure 1-1b. In this drawing, magnets A and B can both rotate freely on the same axis. When magnet A is turned, magnet B follows, and vice-versa, because of magnetic attraction between the two.

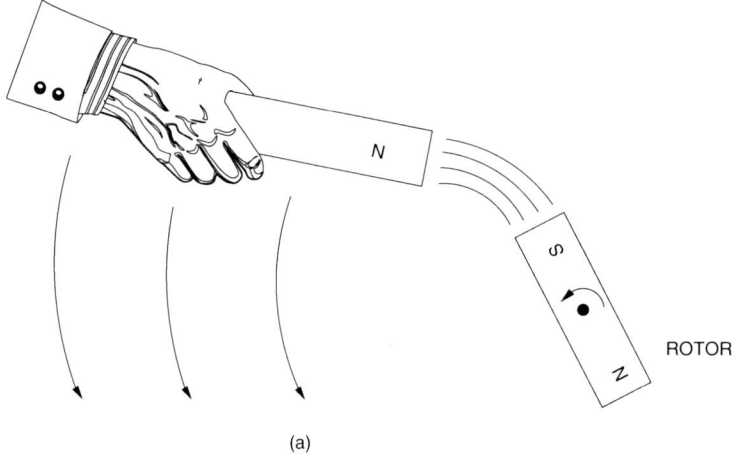

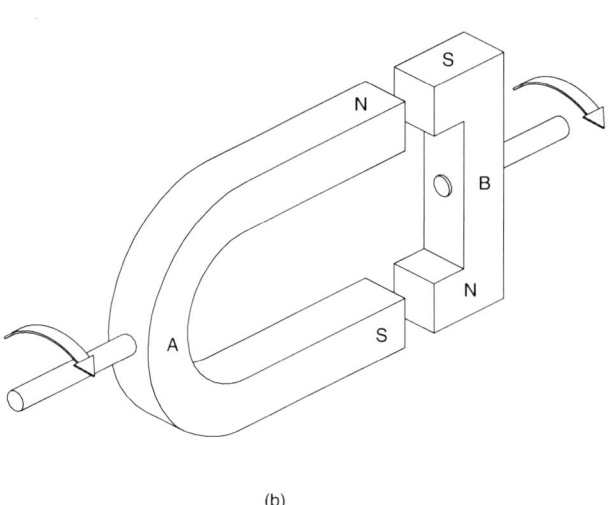

Figure 1-1. Interacting magnetic forces cause motor rotation.

The rotating electromagnet principle

Figure 1-2a shows how magnet A from Figure 1-1b can be replaced with an **electromagnet**. A coil of wire is wrapped around an iron core. The ends of the coil are connected to a dc power source to make current flow in the coil and produce north and south **magnetic poles**. Magnet A is now an electromagnet.

When the electromagnet is rotated manually, magnet B also rotates, as when two magnets were used. At first glance, this setup seems to offer no advantage because a first object (the electromagnet) must still be rotated in order to cause rotation of a second object (magnet B). Furthermore, to prevent the leads interconnecting the dc power source and the electromagnet from twisting, the source must be rotated as the electromagnet rotates, which is not very convenient.

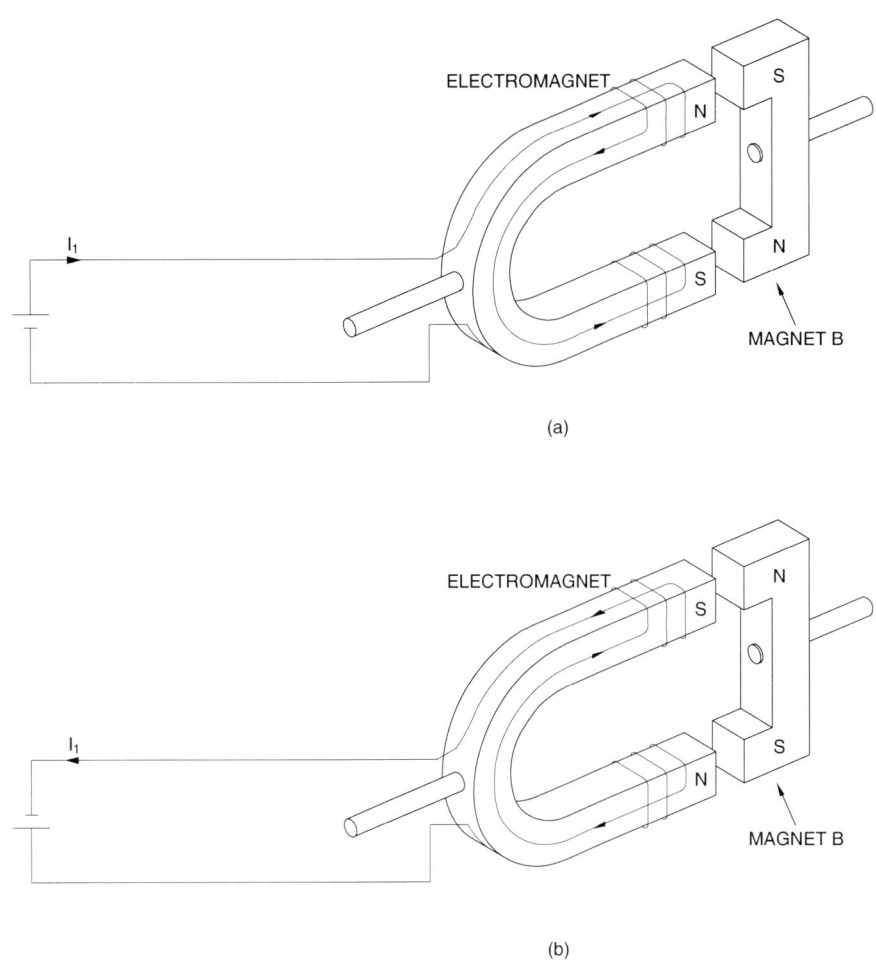

Figure 1-2. Electric current flow produces an electromagnet.

However, if the polarity of the dc power source is reversed as shown in Figure 1-2b, the positions of the north and south poles on the electromagnet are interchanged and magnet B rotates one half turn. Therefore, this setup allows magnet B to be rotated without having to rotate the electromagnet, the direction of rotation being undetermined. By combining two electromagnets, two dc power sources, and changing the voltages and polarities of the sources, it is possible to make magnet B rotates in a given direction without having to move any other objects (electromagnets). Figure 1-3 shows how the electromagnet setup of Figure 1-2 can be modified to achieve this goal. When the currents in the two electromagnets (I_1 and I_2) are as shown in Figure 1-3, the magnetic poles change polarity in sequence at the proper moments, and rotation in the clockwise direction takes place because of the repeating attraction-repulsion process. Therefore, simple current switching has resulted in the electrical equivalent of a rotating magnet. The operating principle of all motors is based on producing the electrical equivalent of a rotating magnet to avoid manual rotation.

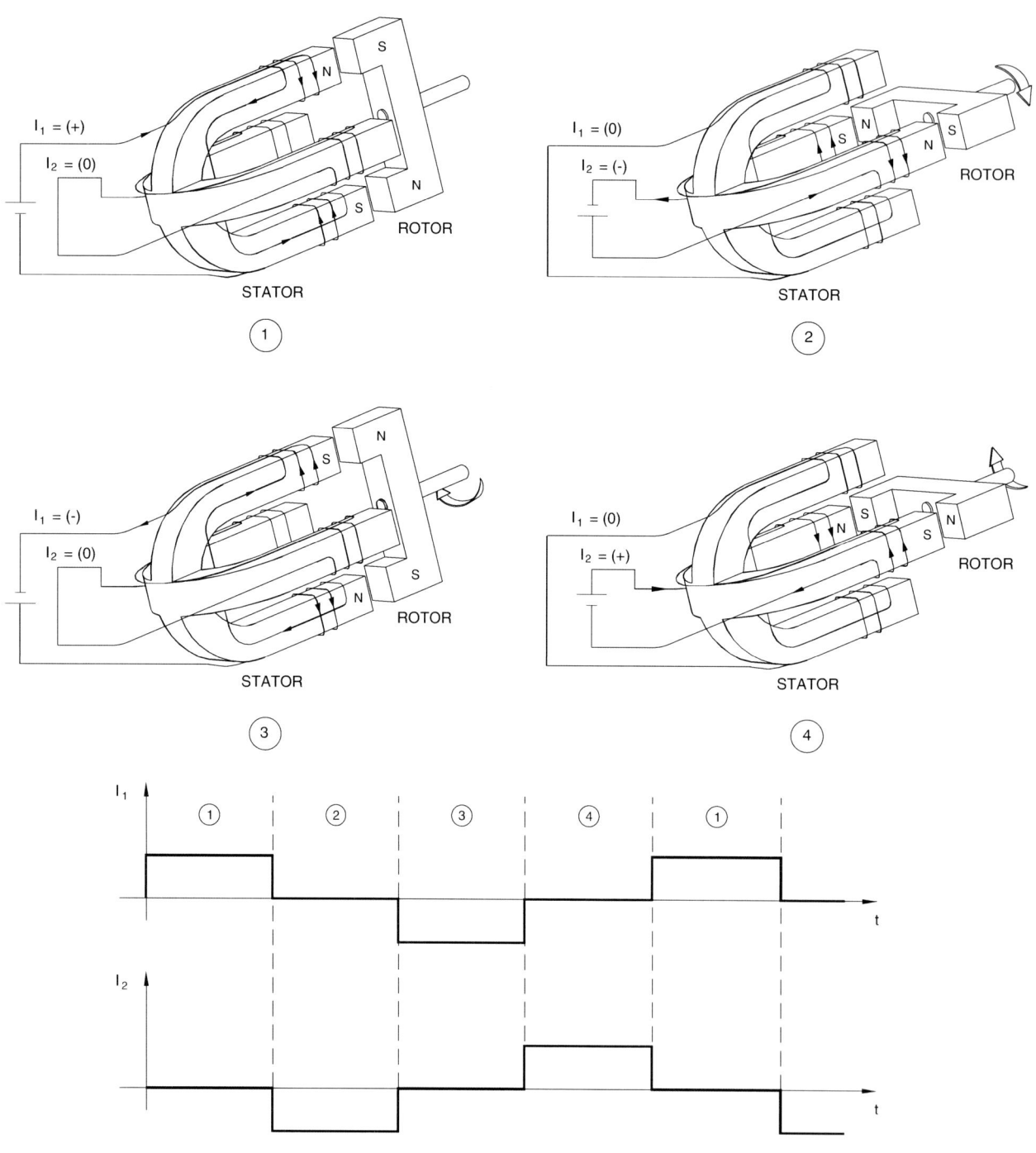

Figure 1-3. Clockwise rotating magnet implemented with an electromagnet.

The direction of rotation can be reversed by interchanging currents I_1 and I_2 as shown in Figure 1-4.

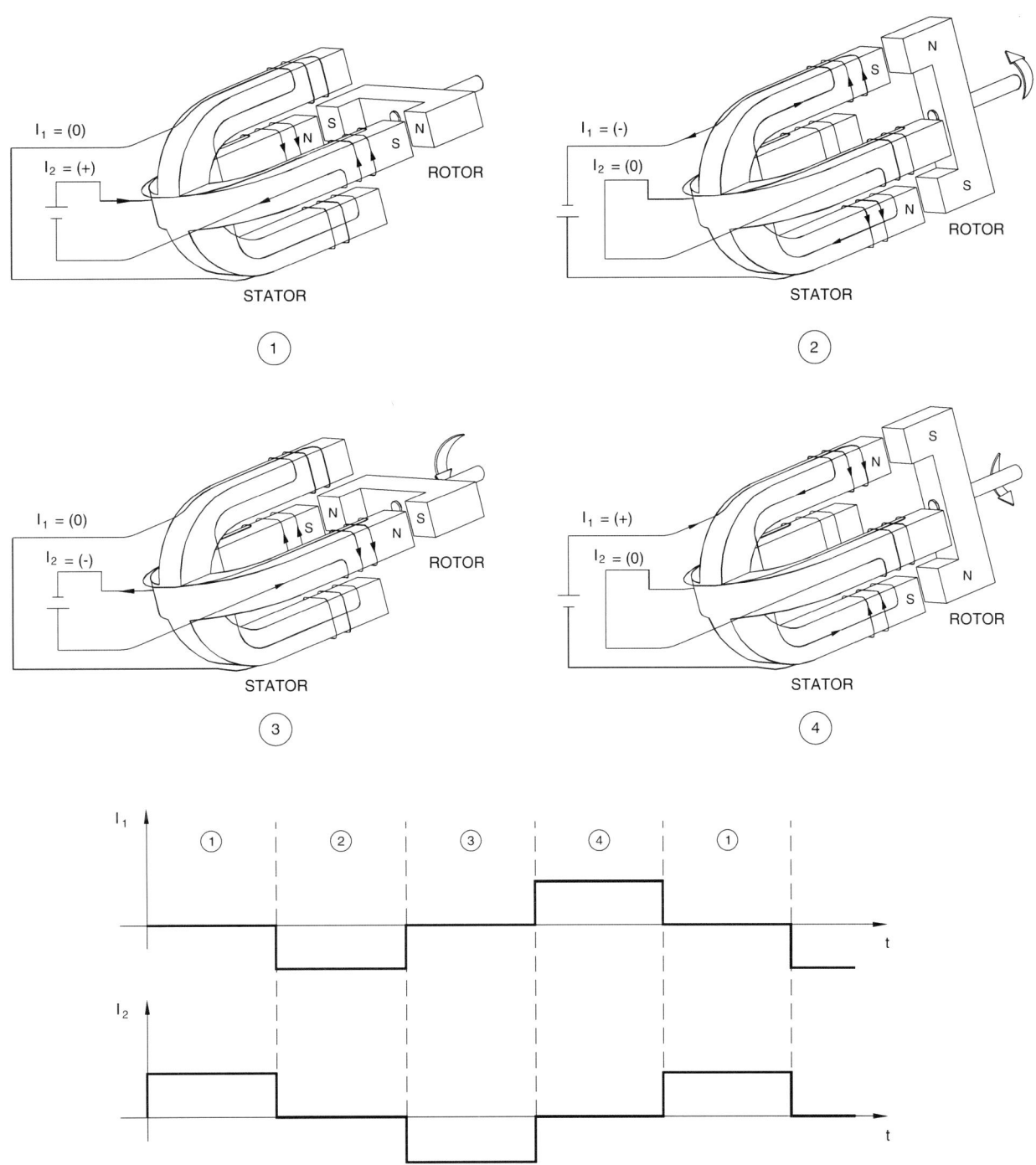

Figure 1-4. Counterclockwise rotating magnet implemented with an electromagnet.

The generator principle

The operation of **generators** and alternators is based on Faraday's law of **electromagnetic induction**, which states the following:

1. A voltage is induced between the terminals of a wire loop if the magnetic flux linking the loop varies as a function of time.

2. The value of the induced voltage is proportional to the rate of change of the magnetic flux.

The induced voltage is given by the following equation, and Figure 1-5 illustrates the principle of magnetic flux that varies inside a coil of N turns.

$$E = N \times \frac{\Delta \varphi}{\Delta t}$$

where E is the induced voltage, in volts (V).
          N is the number of turns of wire in the coil.
          $\Delta \varphi$ is the change of flux inside the coil, in Webers (Wb).
          Δt is the time interval within which the flux changes, in seconds (s).

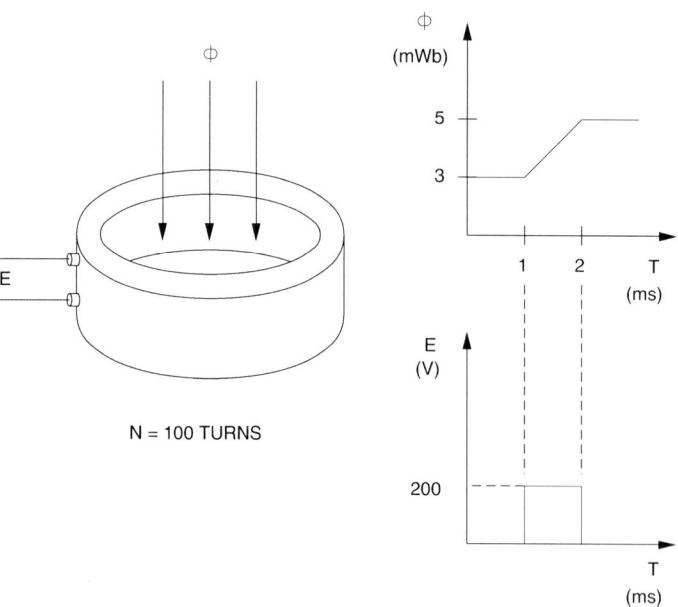

Figure 1-5. Voltage induced in a coil.

For the values given in the circuit of Figure 1-5, the induced voltage will be:

$$E = N \times \frac{\Delta \varphi}{\Delta t} = 100 \times \frac{0.005 - 0.003}{0.001} = 200 \text{ V}$$

If the rotor magnet of Figure 1-3 and Figure 1-4 is turned manually, its magnetic lines of force will induce a voltage in the electromagnet coils, which are usually referred to as the **stator** coils because they never rotate. This will cause current to flow in the stator coils when their ends are short-circuited, thus producing another magnetic field. The interaction between the magnetic fields of the stator coils and the rotor magnet will create a force opposing rotation of the rotor magnet. This is the principle used by a **brake** to create braking torque.

Work, torque, and power

A mechanical work W is done whenever a force F moves an object over a distance d, and work is defined by the following equation:

$$W = F \times d$$

The work W is expressed in joules (J) when the force F and the distance d are expressed in Newtons (N) and in meters (m), respectively. On the other hand, when the force F and the distance d are expressed in pound forces (lbf) and in inches (in), respectively, the work W is expressed in pound force-inches (lbf·in).

As an example, Figure 1-6 shows a block that is moved over a distance of 1 m (39.37 in) by an applied force of 1 N (0.2248 lbf), meaning that 1 J (8.85 lbf·in) of work has been done.

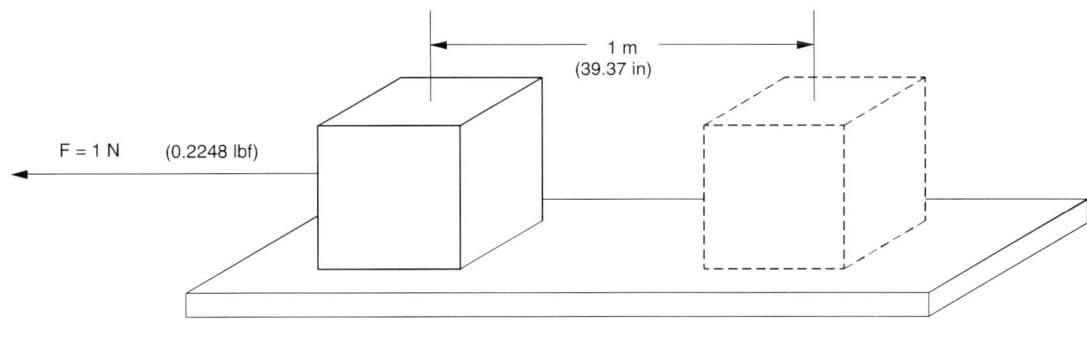

Figure 1-6. Work involved in moving a block.

Now consider that the block is moved over the same distance using a pulley that has a radius r, as shown in Figure 1-7. A twisting force must be applied to turn the pulley so that the rope pulls the block with a force F. This twisting force is known as **torque** and it is defined by the following equation:

$$T = F \times r$$

The torque T is expressed in Newton-meters (N·m) when the force F and the radius r are expressed in Newtons (N) and in meters (m), respectively. On the other hand, when the force F and the radius r are expressed in pound forces (lbf) and in inches (in), respectively, the torque T is expressed in pound force-inches (lbf·in).

At the end of each complete rotation of the pulley, the block has advanced a distance of $2\pi r$ meters (inches), meaning that $2\pi r F$ joules (pound force-inches) of work has been done. Since torque equals $F \times r$, the work may be expressed as $2\pi T$ joules (pound force-inches) per revolution.

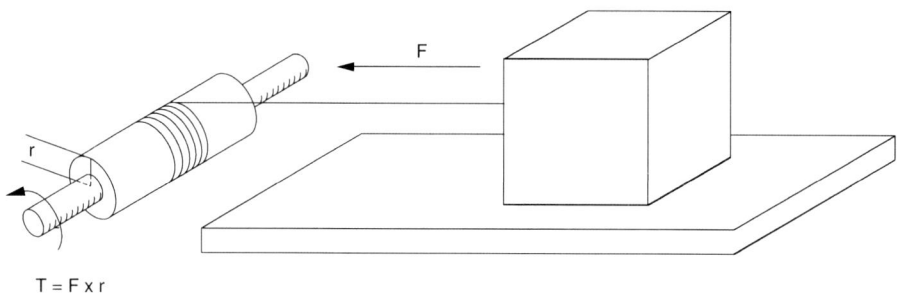

Figure 1-7. Using a pulley to move an object.

Power P is defined as the rate of doing work, and it is given by the following equation when the work W is expressed in joules:

$$P = \frac{W}{t}$$

where P is the power, in watts (W).

            W is the work, in joules (J).

            t is the time taken to do the work, in seconds (s).

When work W is expressed in pound force-inches, the following equation must be used to calculate the power P:

$$P = \frac{W}{8.85 \times t}$$

Since power is work done per unit of time, the power of a motor turning at a **speed** n, can be found using the following equation when the torque T is expressed in newton-meters:

$$P = 2\pi T \times n \times \frac{1}{60}$$

 In this equation, the term 1/60 is used to convert the speed n, expressed in r/min, into a speed expressed in revolutions per second (r/s).

This equation can be simplified as follows:

$$P = \frac{n \times T}{9.55}$$

When torque T is expressed in pound force-inches, the power of the motor can be found using the following equation:

$$P = \frac{2\pi T}{8.85} \times n \times \frac{1}{60}$$

This equation can be simplified as follows:

$$P = \frac{n \times T}{84.51}$$

Exercise 1-1

Prime Mover and Brake Operation (Model 8960-2)

 If you are using the Prime Mover/Dynamometer, model 8960-1, skip this exercise and perform Exercise 1-2 instead.

EXERCISE OBJECTIVE

When you have completed this exercise, you will be able to demonstrate the operation of a prime mover and a brake, using the Four-Quadrant Dynamometer/Power Supply, Model 8960-2. You will be able to measure the opposition torque caused by a machine driven by a dynamometer working as a prime mover. You will be able to measure the output torque of a drive motor using a dynamometer working as a brake.

DISCUSSION

Prime mover operation

The **prime mover** in the Four-Quadrant Dynamometer/Power Supply operates basically like a linear voltage-to-speed converter as illustrated in Figure 1-8, and the direction of rotation is directly related to the input voltage polarity. A positive voltage produces clockwise rotation, while reversing the input voltage polarity results in counterclockwise or negative rotation. The speed-voltage relationship is a straight line, and the higher the applied voltage, the faster the motor turns. The prime mover uses a dc motor, which will be seen in Unit 2.

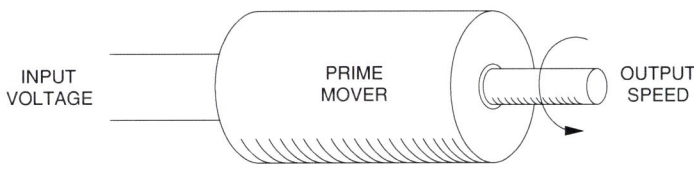

Figure 1-8. The prime mover is a voltage-to-speed converter.

The direction of rotation of the prime mover is selected using the FUNCTION button on the Four-Quadrant Dynamometer/Power Supply. The rotation speed of the prime mover is set by using a speed control (COMMAND knob) on the Four-Quadrant Dynamometer/Power Supply. The prime mover speed can be set to any value between 0 and 2500 r/min.

A digital display on the Four-Quadrant Dynamometer/Power Supply allows direct reading of both the speed and torque values. Connecting ANALOG OUTPUTS T (torque) and n (speed) of the Four-Quadrant Dynamometer/Power Supply to the corresponding inputs on the data acquisition module permits measurement and display of speed and torque using the Speed and Torque meters in the Metering window of the software. The prime mover thus not only acts as a prime mover but also as a speed meter (tachometer) and a dynamometer.

The displayed speed, either on the module display or the Speed meter in the Metering window, is the actual speed at which the prime mover rotates. It is positive for clockwise (CW) rotation and negative for counterclockwise (CCW) rotation.

To rotate, the prime mover must produce sufficient **magnetic torque** ($T_M(P.M.)$) to overcome all forces that oppose its rotation. The combined effect of all these forces results in a torque that opposes the prime mover rotation. This torque is known as the opposition torque ($T_{OPP.}$). As a result, when the prime mover rotates at constant speed, the magnetic torque $T_M(P.M.)$ and the opposition torque $T_{OPP.}$ are equal in magnitude but are of opposite polarity, i.e., $T_M(P.M.) = -T_{OPP.}$

The opposition torque ($T_{OPP.}$) is displayed on the Four-Quadrant Dynamometer/Power Supply and the Torque meter in the Metering window. Therefore, the displayed torque for clockwise (positive) rotation is negative. For counterclockwise (negative) rotation, the displayed torque is positive because the forces opposing rotation always act in the opposite direction. In other words, the prime mover torque and speed displayed are always of opposite polarity.

When no rotating machine is coupled to the prime mover's shaft, the opposition to rotation is only due to the bearing friction, windage friction, and **brushes** friction in the prime mover. The combined effect of these frictions results in the prime mover friction torque $T_F(P.M.)$, as indicated in Figure 1-10 and the following equation:

$$T_F(P.M.) = T_{BRUSHES} + T_{BEARING} + T_{WINDAGE}$$

where $T_{BRUSHES}$ is the torque that opposes rotation which results from the brushes friction.

          $T_{BEARING}$ is the torque that opposes rotation which results from the bearing friction.

          $T_{WINDAGE}$ is the torque that opposes rotation which results from the windage friction.

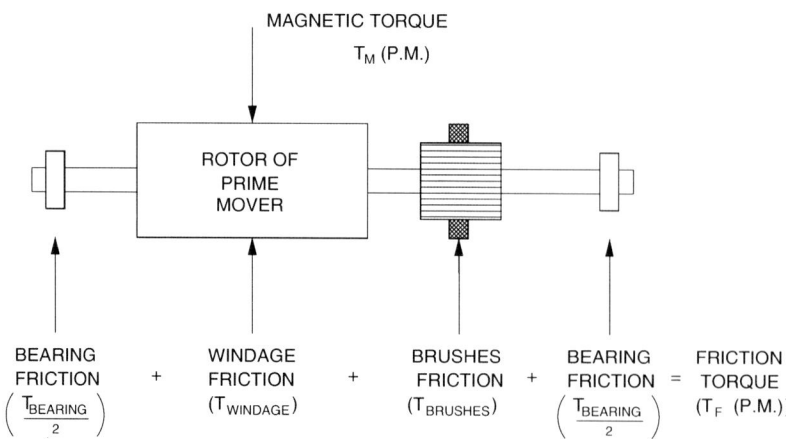

Figure 1-9. Distribution of torque in the prime mover.

Ex. 1-1 – Prime Mover and Brake Operation (Model 8960-2) ♦ *Discussion*

When no rotating machine is coupled to the prime mover's shaft, the prime mover friction torque $T_F(P.M.)$ is the only opposition to prime mover rotation, and therefore, the opposition torque $T_{OPP.}$ is equal to the prime mover friction torque $T_F(P.M.)$. Note that the prime mover friction torque $T_F(P.M.)$, and thereby, the opposition torque $T_{OPP.}$, increase as speed increases. However, this torque-versus-speed relationship is not linear.

When the prime mover is mechanically coupled to another rotating machine, the opposition torque $T_{OPP.}$ increases because of additional opposition to rotation caused by the other machine and the coupling device (timing belt), as indicated in the following equation:

$$T_{OPP.} = T_F(P.M.) + T_F(BELT) + T_{OPP.}(MACHINE)$$

where $T_F(P.M.)$ is the prime mover friction torque.

$T_F(BELT)$ is the torque that opposes rotation which results from the belt friction.

$T_{OPP.}(MACHINE)$ is the torque that opposes rotation which is caused by the machine coupled to the prime mover.

Therefore, the opposition torque $T_{OPP.}$ indicated by the display of the Four-Quadrant Dynamometer/Power Supply and the Torque meter in the Metering window is higher when a rotating machine is coupled to the prime mover.

When the prime mover drives another rotating machine, it is often useful to know the opposition to rotation caused by the driven machine. In other words, it is useful to know the torque that opposes the prime mover rotation which is caused by the driven machine ($T_{OPP.}(MACHINE)$). This torque is equal to the opposition torque $T_{OPP.}$ minus the sum of the prime mover friction torque ($T_F(P.M.)$) and belt friction torque ($T_F(BELT)$), as indicated in the following equation:

$$T_{OPP.}(MACHINE) = T_{OPP.} - \big(T_F(P.M.) + T_F(BELT)\big)$$

A function in the Metering window allows the torque indicated by the Torque meter ($T_{OPP.}$) to be corrected so that it corresponds to the opposition torque produced by the driven machine ($T_{OPP.}(MACHINE)$). This function simply subtracts the typical value of $T_F(P.M.)$ and $T_F(BELT)$ from $T_{OPP.}$ to obtain $T_{OPP.}(MACHINE)$. The corrected torque ($T_{OPP.}(MACHINE)$) is always smaller than the original uncorrected torque ($T_{OPP.}$).

Brake operation

The brake in the Four-Quadrant Dynamometer/Power Supply is a device which can be coupled to a drive motor using a timing belt. It is used to mechanically load the motor. Furthermore, the brake also measures the speed and **output torque** of the motor. In other words, it acts as a brake as well as a speed meter and dynamometer, all at the same time. The mechanical load to the motor can be changed using a torque control (COMMAND knob) on the Four-Quadrant Dynamometer/Power Supply. This allows study of the behaviour of a motor under various load conditions.

As for the prime mover, the display on the Four-Quadrant Dynamometer/Power Supply allows direct reading of the speed and torque values. Connecting ANALOG OUTPUTS T (torque) and n (speed) of the Four-Quadrant Dynamometer/Power Supply to the corresponding inputs on the data acquisition module permits measurement and display of speed and torque using the Speed and Torque meters in the Metering window of the software.

The displayed speed, either on the module display or the Speed meter in the Metering window, is the actual speed at which the brake rotates. It is positive for clockwise rotation and negative for counterclockwise rotation.

The mechanical load which the brake produces when it is coupled to a drive motor consists of friction in the coupling device (timing belt), friction in the brake (bearing, brushes, and windage frictions), and a magnetic torque which the brake produces to oppose rotation of the motor ($T_M(BRK.)$). The combined effect of these frictions and this torque results in a load torque (T_{LOAD}) that opposes rotation of the motor coupled to the brake, as indicated in the following equation:

$$T_{LOAD} = T_F(BELT) + T_F(BRK.) + T_M(BRK.)$$

where $T_F(BELT)$ is the torque that opposes rotation which results from the belt friction.

$T_F(BRK.)$ is the torque that opposes rotation which results from friction in the brake.

$T_M(BRK.)$ is the magnetic torque produced in the brake to oppose rotation.

Figure 1-10 illustrates the forces that oppose rotation when a drive motor is coupled to the brake.

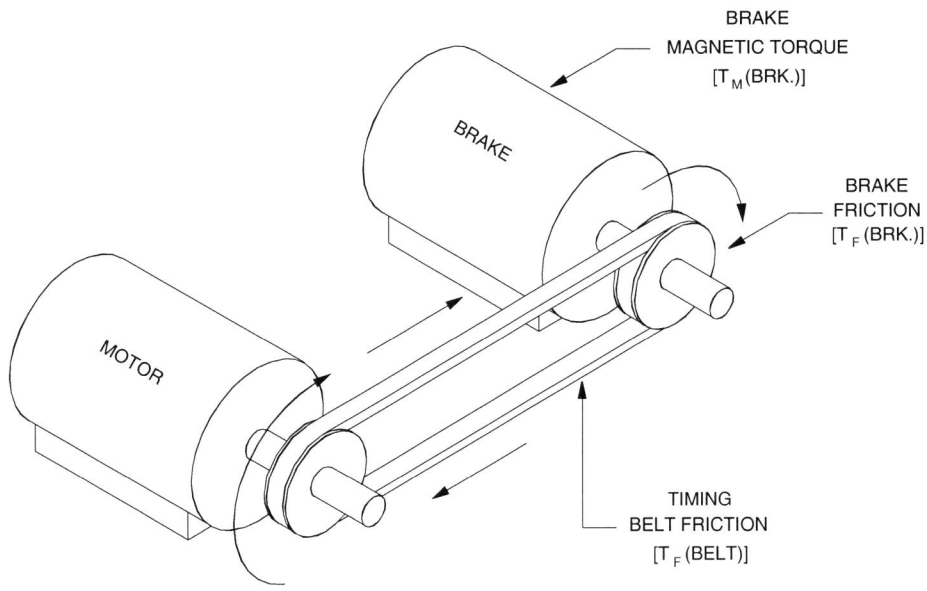

Figure 1-10. Forces that oppose motor rotation.

Torques $T_F(BELT)$ and $T_F(BRK.)$, which result from friction, vary as speed varies, as is the case with torque $T_F(P.M.)$ seen in the previous subsection. On the other hand, torque $T_M(BRK.)$ does not vary with speed, but it can be varied over a range of 0 to 3 N·m (0 to 27 lbf·in) using the torque control of the brake. Therefore, the load torque T_{LOAD} can be varied by varying torque $T_M(BRK.)$.

The torque indicated by the display of the Four-Quadrant Dynamometer/Power Supply and the Torque meter in the Metering window has the same magnitude as the magnetic torque $T_M(BRK.)$ but is of opposite polarity. In other words, the indicated torque is equal to $-T_M(BRK.)$. This means that the torque indicated by the display of the Four-Quadrant Dynamometer/Power Supply and the Torque meter is equal to the torque which the motor coupled to the brake must provide to overcome the opposition to rotation caused by torque $T_M(BRK.)$. Therefore, the displayed torque for clockwise (positive) rotation is positive. For counterclockwise (negative) rotation, the displayed torque is negative. In brief, the brake torque and speed displayed are always of the same polarity. Furthermore, when the COMMAND knob is set to minimum [fully counterclockwise, or CCW position], torque $T_M(BRK.)$ is zero, and therefore, the torque indicated is also 0. This torque increases as the COMMAND knob is turned clockwise.

Ex. 1-1 – Prime Mover and Brake Operation (Model 8960-2) ◆ *Discussion*

However, the torque indicated by the display of the Four-Quadrant Dynamometer/Power Supply and the Torque meter (i.e., the magnetic torque) does not correspond to the output torque (T_{OUT}) of the motor coupled to the brake. To rotate, this motor must produce sufficient output torque T_{OUT} to overcome the load torque T_{LOAD}, which consists of the brake magnetic torque $T_M(BRK.)$ plus the belt friction torque $T_F(BELT)$ plus the brake friction torque $T_F(BRK.)$. In other words, the motor output torque T_{OUT} must be equal to the load torque T_{LOAD} but of opposite polarity. This is indicated in the following equation:

$$T_{OUT} = T_{LOAD} = -\bigl(T_M(BRK.) + T_F(BELT) + T_F(BRK.)\bigr)$$

A function allows the torque indicated by the Torque meter in the Metering window to be corrected so that it indicates the motor output torque T_{OUT}. This function simply adds the typical value of $T_F(BELT)$ and $T_F(BRK.)$ to $T_M(BRK.)$. The corrected torque, i.e., the motor output torque T_{OUT}, is thus always greater than the original uncorrected torque.

Procedure summary

In the first part of this exercise, you will experiment with the Four-Quadrant Dynamometer/Power Supply operating in the prime mover mode. You will set up the equipment in the Workstation, connect the equipment as shown in Figure 1-11, and make the appropriate settings on the Four-Quadrant Dynamometer/Power Supply. You will learn how to change the direction of rotation of the prime mover by using the FUNCTION button on the Four-Quadrant Dynamometer/Power Supply. You will check that the polarity of the speed indicated on the Speed meter in the Metering window changes when the direction of rotation is changed. You will then measure the prime mover friction torque. Finally, you will couple the prime mover to a squirrel-cage induction motor and measure the opposition torque caused by the motor.

In the second part of this exercise, you will experiment with the Four-Quadrant Dynamometer/Power Supply operating in the brake mode. You will couple the brake to a dc motor. You will vary the setting of the torque command (COMMAND knob) of the brake while observing the speed and torque polarity indicated by the Speed and Torque meters in the Metering window. You will learn how to correct the torque indicated by the Torque meter to determine the output torque of the dc motor.

EQUIPMENT REQUIRED

Refer to the Equipment Utilization Chart in Appendix C to obtain the list of equipment required for this exercise.

PROCEDURE

 High voltages are present in this laboratory exercise. Do not make or modify any banana jack connections with the power on unless otherwise specified.

Setting up the equipment

 If you are performing this exercise using LVSIM-EMS, skip manipulation 1 and 2.

1. Install the equipment required in the EMS workstation.

2. On the Power Supply, make sure the main power switch is set to the O (off) position, and the voltage control knob is turned fully counterclockwise. Ensure the Power Supply is connected to a three-phase power source.

3. Ensure that the data acquisition module is connected to a USB port of the host computer.

   Connect the POWER INPUT of the data acquisition module to the 24 V - AC output of the Power Supply.

   On the Power Supply, set the 24 V - AC power switch to the I (on) position.

4. Start the Data Acquisition software (LVDAC or LVDAM). Open setup configuration file DCMOTOR1.DAI.

    If you are using LVSIM-EMS in LVVL, you must use the IMPORT option in the File menu to open the configuration file.

   In the Metering window, select layout 1 via the View or Options menu. Make sure that the continuous refresh mode is selected.

    *If you are using LVDAC-EMS, set parameter **Analog Input AI7** in the Data Acquisition and Control Settings window to **Non-Corrected Torque (N·m)**.*

5. Connect the equipment as shown in Figure 1-11.

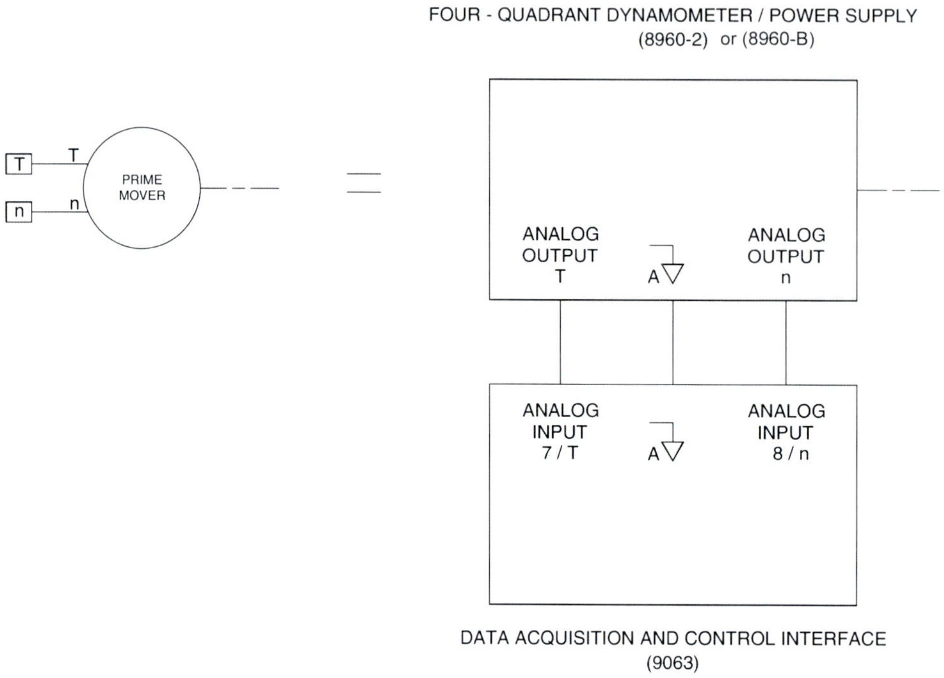

Figure 1-11. Prime mover circuit.

6. Connect the POWER INPUT of the Four-Quadrant Dynamometer/Power Supply to a wall receptacle.

   Turn the Four-Quadrant Dynamometer/Power Supply on by setting the POWER INPUT switch to the I (on) position.

   On the Four-Quadrant Dynamometer/Power Supply, set the OPERATING MODE switch to the DYNAMOMETER position. This setting allows the Four-Quadrant Dynamometer/Power Supply to operate as a brake or a prime mover, depending on the selected function.

   Set the Four-Quadrant Dynamometer/Power Supply to operate as a clockwise prime mover. To do this, momentarily press the FUNCTION button until the function indicated by the display of the Four-Quadrant Dynamometer/Power Supply is CW Prime Mover/Brake.

   On the Four-Quadrant Dynamometer/Power Supply, press and hold the FUNCTION button 3 seconds to have uncorrected torque values on the display. The indication "NC" appears next to the function name on the display when uncorrected torque values are indicated.

   By default, the torque correction function is enabled in the Four-Quadrant Dynamometer/Power Supply. This function can be disabled by pressing and holding the FUNCTION button 3 seconds. The torque correction function can be enabled again by pressing the FUNCTION button once again for 3 seconds. The status (enabled or disabled) of the torque correction function stays unchanged when another function is selected with the FUNCTION button.

Speed, polarity and direction of rotation

7. On the Four-Quadrant Dynamometer/Power Supply, set the speed command of the prime mover to about 250 r/min, using the COMMAND knob. The value of the speed command is indicated by the module display. Notice that the speed command shown on the module display is blinking.

   On the Four-Quadrant Dynamometer/Power Supply, start the prime mover by momentarily pressing the START/STOP button. Observe that the prime mover starts to rotate. Also notice that the speed value on the module display is no longer blinking to indicate that the indicated speed is now the actual rotation speed of the prime mover.

   In the following blank space, record the speed n of the primer mover indicated by the display of the Four-Quadrant Dynamometer/Power Supply.

   n (prime mover speed) = _____ r/min

   Notice that the Speed meter in the Metering window also indicates the prime mover speed.

   Observe the prime mover. What is the direction of rotation?

   Direction of rotation: _____

8. On the Four-Quadrant Dynamometer/Power Supply, stop the prime mover by momentarily pressing the START/STOP button.

   Set the Four-Quadrant Dynamometer/Power Supply to operate as a counterclockwise prime mover. To do this, momentarily press the FUNCTION button until the function indicated by the display of the Four-Quadrant Dynamometer/Power Supply is CCW Prime Mover/Brake.

   Set the speed command of the prime mover to about -250 r/min using the COMMAND knob. Start the prime mover by momentarily pressing the START/STOP button.

   Observe the prime mover. What is the direction of rotation?

   Direction of rotation: _____

   With the CCW Prime Mover/Brake function, what difference is there in the speed indicated by the Speed meter in the Metering window?

   _____

   _____

 In this manual, clockwise rotation is defined as the positive direction, and counterclockwise rotation is therefore indicated by a negative speed value. The speed indicated by the display of the Four-Quadrant Dynamometer/Power Supply and the Speed meter in the Metering window uses this same convention.

Ex. 1-1 – Prime Mover and Brake Operation (Model 8960-2) ♦ *Procedure*

Measuring the prime mover friction torque

9. On the Four-Quadrant Dynamometer/Power Supply, stop the prime mover by momentarily pressing the START/STOP button.

   Select the clockwise prime mover function by momentarily pressing the FUNCTION button until the function indicated by the display of the Four-Quadrant Dynamometer/Power Supply is CW Prime Mover/Brake.

   Set the speed command of the prime mover to about 1500 r/min using the COMMAND knob. Start the prime mover by momentarily pressing the START/STOP button.

10. Record the friction torque $T_F(P.M.)$ indicated by the display of the Four-Quadrant Dynamometer/Power Supply.

    $T_F(P.M.) = $ _____ N·m (lbf·in)

    Why is the torque indicated by the module display negative while the prime mover speed is positive (clockwise rotation)?

    _____

    _____

    Notice that the Torque meter in the Metering window indicates approximately the same torque as the display of the Four-Quadrant Dynamometer/Power Supply. The torque indicated by the Torque meter is not corrected, since the indication "NC" appears in the lower left corner of this meter.

11. On the Four-Quadrant Dynamometer/Power Supply, stop the prime mover by momentarily pressing the START/STOP button.

    Select the counterclockwise prime mover function by momentarily pressing the FUNCTION button until the function indicated by the display of the Four-Quadrant Dynamometer/Power Supply is CCW Prime Mover/Brake.

    Set the speed command of the prime mover to -1500 r/min using the COMMAND knob. Start the prime mover by momentarily pressing the START/STOP button.

Is the torque indicated by the Torque meter in the Metering window ($T_F(P.M.)$) of opposite sign but approximately equal to the value recorded for the same speed?

❏ Yes   ❏ No

Measuring the opposition torque caused by the driven machine

12. Stop the prime mover by setting the POWER INPUT switch of the Four-Quadrant Dynamometer/Power Supply to the O (off) position.

    Use a timing belt to mechanically couple the Four-Quadrant Dynamometer/Power Supply to the Four-Pole Squirrel-Cage Induction Motor.

    Before installing or removing a timing belt, make absolutely sure that power is turned off to prevent any rotating machine from starting inadvertently.

13. Turn the Four-Quadrant Dynamometer/Power Supply on by setting the POWER INPUT switch to the I (on) position.

    Select the clockwise prime mover function by momentarily pressing the FUNCTION button until the function indicated by the display of the Four-Quadrant Dynamometer/Power Supply is CW Prime Mover/Brake.

    Press and hold the FUNCTION button 3 seconds to have uncorrected torque values on the display of the Four-Quadrant Dynamometer/Power Supply. The indication "NC" appears next to the function name on the display when uncorrected torque values are indicated.

    Set the speed command of the prime mover to about 1500 r/min using the COMMAND knob. Start the prime mover by momentarily pressing the START/STOP button.

    Record the opposition torque ($T_{OPP.}$) indicated by the Torque meter in the Metering window.

    $T_{OPP.} = $ _____ N·m (lbf·in)

    Compare the opposition torque measured in this step to the opposition torque ($T_F(P.M.)$) recorded in step 10. Briefly explain the difference between these two opposition torques.

    _____
    _____
    _____

14. In the Metering window, enable the torque correction function for the Torque meter by clicking the button in the lower left corner of the meter. The indication "C" appears in this button to indicate that the torque correction is enabled. The Torque meter now indicates the opposition torque caused by the Four-Pole Squirrel-Cage Induction Motor $T_{OPP.}(MACHINE)$. Record this torque in the following blank space.

$T_{OPP.}(MACHINE) =$ _____ N·m (lbf·in)

Using the torque measured in this step and the previous step, compare the opposition torque caused by the Four-Pole Squirrel-Cage Induction Motor ($T_{OPP.}(MACHINE)$) to the total opposition torque ($T_{OPP.}$).

_____
_____
_____

15. Stop the prime mover by setting the POWER INPUT switch of the Four-Quadrant Dynamometer/Power Supply to the O (off) position.

On the Four-Pole Squirrel-Cage Induction Motor, tilt the front panel forward to get access to the motor's shaft.

Turn the motor's shaft manually. While doing this, notice the twisting force you must apply to make the motor's shaft rotate.

⚠ WARNING

Before turning a motor's shaft manually or removing a timing belt, make absolutely sure that power is turned off to prevent any rotating machine from starting inadvertently.

Remove the timing belt that couples the Four-Quadrant Dynamometer/Power Supply to the Four-Pole Squirrel-Cage Induction Motor.

Turn the motor's shaft manually. While doing this, notice the twisting force you must apply to make the motor's shaft rotate. Is this force much smaller than that required to make the motor's shaft rotate when the motor is coupled to the prime mover?

❑ Yes ❑ No

Is your observation the same as when you compared the opposition torques in step 13? Briefly explain.

_____
_____
_____

Brake operation

16. Remove the Four-Pole Squirrel-Cage Induction Motor from the EMS Workstation. Install a DC Motor/Generator in the EMS workstation (next to the Four-Quadrant Dynamometer/Power Supply).

Use a timing belt to mechanically couple the Four-Quadrant Dynamometer/Power Supply to the DC Motor/Generator.

Connect the equipment as shown in Figure 1-12.

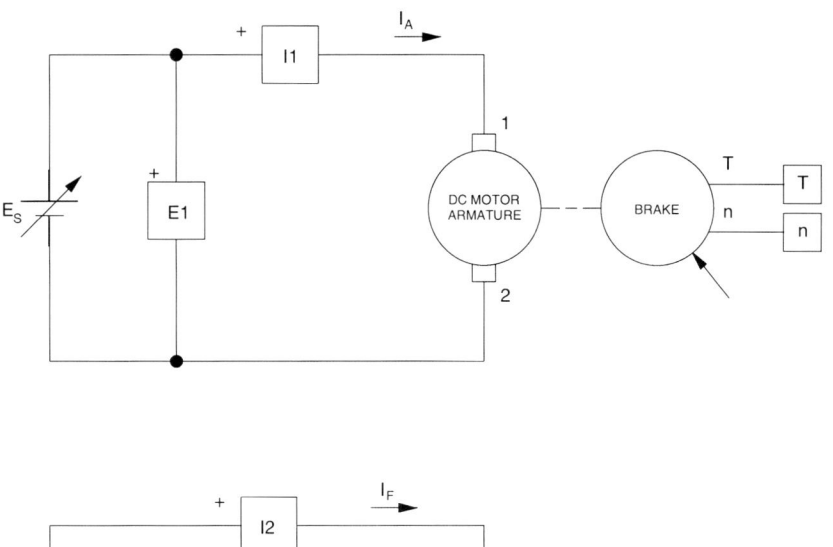

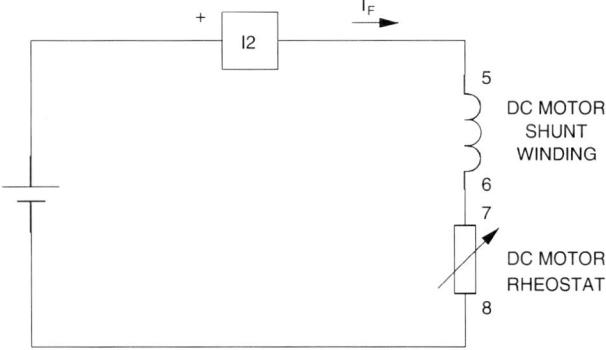

Figure 1-12. DC motor coupled to a brake.

17. Turn the Four-Quadrant Dynamometer/Power Supply on by setting the POWER INPUT switch to the I (on) position.

Set the Four-Quadrant Dynamometer/Power Supply to operate as a brake. To do this, momentarily press the FUNCTION button until the function indicated by the display is 2Q CT Brake.

Make sure that the display of the Four-Quadrant Dynamometer/Power Supply indicates uncorrected torque values (i.e., the indication "NC" appears next to the function name on the display).

Set the torque command of the brake to minimum by setting the COMMAND knob fully counterclockwise.

Start the Four-Quadrant Dynamometer/Power supply by momentarily pressing the START/STOP button.

18. Turn the Power Supply on.

    On the DC Motor/Generator, set the FIELD RHEOSTAT so that the current indicated by meter I field (I_f) in the Metering window is equal to the value given in Table 1-1.

    Table 1-1. DC motor field current.

Local ac power network		I_f (mA)
Voltage (V)	Frequency (Hz)	
120	60	300
220	50	190
220	60	190
240	50	210

    On the Power Supply, set the voltage control knob so that the dc motor rotates at a speed of 1500 r/min. The display of the Four-Quadrant Dynamometer/Power Supply and the Speed meter in the Metering window indicate the motor speed.

    In the Metering window, disable the torque correction function for the Torque meter by clicking the button in the lower left corner of the meter. The indication "NC" appears in this button when the torque correction is disabled.

    Observe that the torque indicated by the display of the Four-Quadrant Dynamometer/Power Supply and the Torque meter in the Metering window is zero. Briefly explain why.

    _____

    _____

19. On the Four-Quadrant Dynamometer/Power Supply, slowly turn the COMMAND knob (torque control) clockwise until the torque indicated by the module display and the Torque meter in the Metering window is equal to 1.0 N·m (9.0 lbf·in). While doing this, observe the speed indicated by the Speed meter in the Metering window.

    What happens to the speed as the torque passes from 0 to 1.0 N·m (0 to 9.0 lbf·in)?

    _____

    _____

Notice that the torque has the same polarity as the speed n. Why?

___

___

___

Measuring the output torque of a drive motor

20. On the Power Supply, readjust the voltage control knob so that the dc motor rotates at a speed of 1500 r/min.

    Make sure that the torque indicated by the display of the Four-Quadrant Dynamometer/Power Supply and the Torque meter in the Metering window is equal to 1.0 N·m (9.0 lbf·in). Slightly readjust the torque control (COMMAND knob) if necessary.

    Record the dc motor torque indicated by the Torque meter in the following blank space. Explain.

    $T_{OUT}(\text{uncorrected}) = $ _____ N·m (lbf·in)

    $[n = 1500 \text{ r/min}]$

    $[T_M(BRK.) = 1 \text{ N·m } (9.0 \text{ lbf·in})]$

21. In the Metering window, enable the torque correction function for the Torque meter by clicking the button in the lower left corner of the meter. The indication "C" appears in this button to indicate that the torque correction is enabled. The Torque meter now indicates the dc motor output torque T_{OUT} (corrected). Record this torque in the following blank space.

    $T_{OUT} \text{ (corrected)} = $ _____ N·m (lbf·in)

    $[n = 1500 \text{ r/min}]$

    $[T_M(BRK.) = -1 \text{ N·m } (-9.0 \text{ lbf·in})]$

    What happened to the torque indicated by the Torque meter when you selected the torque correction function? Briefly explain why.

___

___

___

22. On the Power Supply, turn the voltage control knob fully counterclockwise and set the main power switch to the O (off) position.

Reverse the connection of the leads at the armature (terminals 1 and 2) of the DC Motor/Generator.

Turn the Power Supply on and set the voltage control knob so that the dc motor rotates at a speed of -1500 r/min.

Record the dc motor output torque T_{OUT} (corrected) in the following blank space.

T_{OUT} (corrected) = _____ N·m (lbf·in)

[n = -1500 r/min]
[$T_M(BRK.)$ = 1 N·m (9.0 lbf·in)]

What effect does changing the direction of rotation have on the dc motor output torque T_{OUT} (corrected)?

_____

_____

23. On the Power Supply, set the main power switch and the 24 V - AC power switch to the O (off) position.

Turn the Four-Quadrant Dynamometer/Power Supply off by setting its POWER INPUT switch to the O (off) position.

Remove all leads and cables.

CONCLUSION

In the first part of this exercise, you learned that the prime mover in the Four-Quadrant Dynamometer/Power Supply operates basically like a linear voltage-to-speed converter. The prime mover speed and direction of rotation can be changed by using the COMMAND knob and FUNCTION button, respectively, on the Four-Quadrant Dynamometer/Power Supply. The speed indicated on the Speed meter in the Metering window is positive for clockwise rotation, and negative for counterclockwise rotation. The torque indicated on the Torque meter in the Metering window is the torque that opposes the prime mover rotation (opposition torque). A torque correction function can be enabled so that the meter indicates the opposition torque caused by the driven machine.

Ex. 1-1 – Prime Mover and Brake Operation (Model 8960-2) ♦ *Review Questions*

In the second part of this exercise, you learned that the brake in the Four-Quadrant Dynamometer/Power Supply is used to mechanically load a motor. Furthermore, the brake also measures the speed and output torque of the motor. In other words, it acts as a brake as well as a speed meter and dynamometer, all at the same time. You saw that the mechanical load (torque command) to the motor can be changed by using the COMMAND knob on the Four-Quadrant Dynamometer/Power Supply. When the torque command is increased, the motor speed decreases due to the increased mechanical load on the motor. The Torque meter in the Metering window indicates the torque which the motor must produce to overcome the opposition torque produced by the brake. A torque correction function can be enabled so that the meter indicates the output torque of the motor.

REVIEW QUESTIONS

1. In the prime mover mode, a negative speed value on the Speed meter in the Metering window indicates that

    a. the prime mover rotation is in the clockwise direction.
    b. the prime mover rotation is in the counterclockwise direction.
    c. the prime mover is supplying torque to drive a load.
    d. the prime mover is receiving power from a load.

2. In the prime mover mode, why does the Torque meter in the Metering window indicate a torque value even with no mechanical load applied to the machine's shaft?

    a. The display is probably defective.
    b. Because the prime mover must supply torque to overcome friction.
    c. This indicates the electrical power supplied to the prime mover.
    d. This indicates the mechanical power supplied to the prime mover.

3. In the prime mover mode, will the torque indicated by the Torque meter be a negative or positive value for clockwise rotation?

    a. Negative.
    b. Positive.
    c. It depends on how fast the prime mover rotates.
    d. It depends on the value of the applied voltage.

4. In the brake mode, what does the Torque meter indicate when the torque correction function is disabled?

   a. The sum of the brake friction torque $T_F(BRK.)$ and belt friction torque $T_F(BELT)$.
   b. The magnetic torque produced by the brake to oppose rotation ($T_M(BRK.)$).
   c. The torque which the motor coupled to the brake must produce to overcome the magnetic torque $T_M(BRK.)$, i.e., the inverse of $T_M(BRK.)$.
   d. The load torque T_{LOAD} produced by the brake.

5. When the brake is coupled to a drive motor, what does the Torque meter indicate when the torque correction function is enabled?

   a. The load torque T_{LOAD}.
   b. The sum of the belt friction torque $T_F(BELT)$, brake friction torque $T_F(BRK.)$, and brake magnetic torque $T_M(BRK.)$.
   c. The output torque T_{OUT} of the motor coupled to the brake minus the belt friction torque $T_F(BELT)$.
   d. The output torque T_{OUT} of the motor coupled by the brake.

Exercise 1-2

Prime Mover and Brake Operation (Model 8960-1)

If you are using the Four-Quadrant Dynamometer/Power Supply, model 8960-2 (or model 8960-B), skip this exercise.

EXERCISE OBJECTIVE

When you have completed this exercise, you will be able to demonstrate the operation of a prime mover and a brake, using the Prime Mover/Dynamometer, Model 8960-1. You will be able to measure the opposition torque caused by a machine driven by a prime mover using a dynamometer. You will be able to measure the output torque of a drive motor using a brake and a dynamometer.

DISCUSSION

Prime mover operation

The prime mover in the Prime Mover/Dynamometer operates basically like a linear voltage-to-speed converter as illustrated in Figure 1-13, and the direction of rotation is directly related to the input voltage polarity. A positive voltage produces clockwise rotation, while reversing the input voltage polarity results in counterclockwise or negative rotation. The speed-voltage relationship is a straight line, and the higher the applied voltage, the faster the motor turns. The prime mover in the Prime Mover/Dynamometer uses a dc motor, which will be seen in Unit 2.

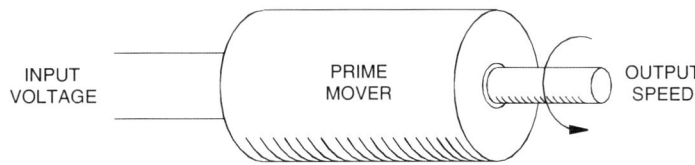

Figure 1-13. The prime mover is a voltage-to-speed converter.

A digital display on the Prime Mover/Dynamometer module allows direct reading of either speed or torque values. Connecting the SPEED and TORQUE OUTPUTS of the Prime Mover/Dynamometer to the corresponding inputs on the data acquisition module permits measurement and display of speed and torque using the Speed and Torque meters in the Metering window of the software. The prime mover thus not only acts as a prime mover but also as a speed meter (tachometer) and a dynamometer.

The displayed speed, either on the module display or the Speed meter in the Metering window, is the actual speed at which the prime mover rotates. It is positive for clockwise (CW) rotation and negative for counterclockwise (CCW) rotation.

Throughout this manual, the abbreviation "CW" stands for clockwise, while the abbreviation "CCW" stands for counterclockwise.

Ex. 1-2 – Prime Mover and Brake Operation (Model 8960-1) ♦ *Discussion*

To rotate, the prime mover must produce sufficient magnetic torque ($T_M(P.M.)$) to overcome all forces that oppose its rotation. The combined effect of all these forces results in a torque that opposes the prime mover rotation. This torque is known as the opposition torque ($T_{OPP.}$). As a result, when the prime mover rotates at constant speed, the magnetic torque $T_M(P.M.)$ and the opposition torque $T_{OPP.}$ are equal in magnitude but are of opposite polarity, i.e., $T_M(P.M.) = -T_{OPP.}$

The opposition torque ($T_{OPP.}$) is displayed on the Four-Quadrant Dynamometer/Power Supply and the Torque meter in the Metering window. Therefore, the displayed torque for clockwise (positive) rotation is negative. For counterclockwise (negative) rotation, the displayed torque is positive because the forces opposing rotation always act in the opposite direction. In other words, the prime mover torque and speed displayed are always of opposite polarity.

When no rotating machine is coupled to the prime mover's shaft, the opposition to rotation is only due to the bearing friction, windage friction, and brushes friction in the prime mover. The combined effect of these frictions results in the prime mover friction torque $T_F(P.M.)$, as indicated in Figure 1-14 and the following equation:

$$T_F(P.M.) = T_{BRUSHES} + T_{BEARING} + T_{WINDAGE}$$

where $T_{BRUSHES}$ is the torque that opposes rotation which results from the brushes friction.

           $T_{BEARING}$ is the torque that opposes rotation which results from the bearing friction.

           $T_{WINDAGE}$ is the torque that opposes rotation which results from the windage friction.

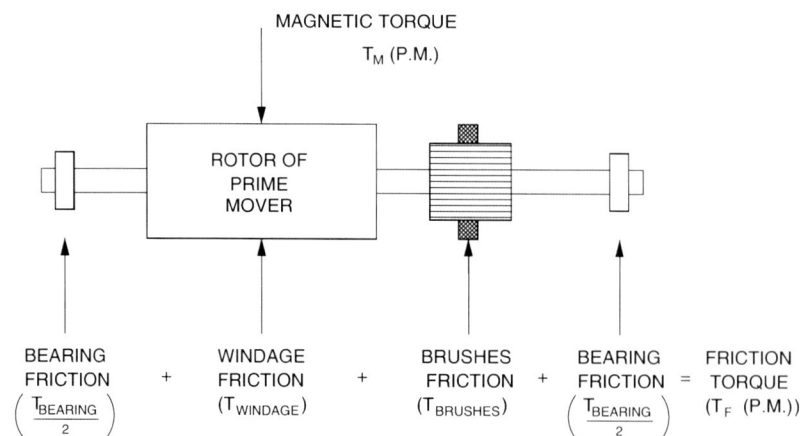

Figure 1-14. Distribution of torque in the prime mover.

When no rotating machine is coupled to the prime mover's shaft, the prime mover friction torque $T_F(P.M.)$ is the only opposition to prime mover rotation, and therefore, the opposition torque $T_{OPP.}$ is equal to the prime mover friction torque $T_F(P.M.)$. Note that the prime mover friction torque $T_F(P.M.)$, and thereby, the opposition torque $T_{OPP.}$, increase as speed increases. However, this torque-versus-speed relationship is not linear.

When the prime mover is mechanically coupled to another rotating machine, the opposition torque $T_{OPP.}$ increases because of additional opposition to rotation caused by the other machine and the coupling device (timing belt), as indicated in the following equation:

$$T_{OPP.} = T_F(P.M.) + T_F(BELT) + T_{OPP.}(MACHINE)$$

where $T_F(P.M.)$ is the prime mover friction torque.

$T_F(BELT)$ is the torque that opposes rotation which results from the belt friction.

$T_{OPP.}(MACHINE)$ is the torque that opposes rotation which is caused by the machine coupled to the prime mover.

Therefore, the opposition torque $T_{OPP.}$ indicated by the module display and on the Torque meter in the Metering window is higher when a rotating machine is coupled to the prime mover.

When the prime mover drives another rotating machine, it is often useful to know the opposition to rotation caused by the driven machine. In other words, it is useful to know the torque that opposes the prime mover rotation which is caused by the driven machine ($T_{OPP.}(MACHINE)$). This torque is equal to the opposition torque $T_{OPP.}$ minus the sum of the prime mover friction torque ($T_F(P.M.)$) and belt friction torque ($T_F(BELT)$), as indicated in the following equation:

$$T_{OPP.}(MACHINE) = T_{OPP.} - \big(T_F(P.M.) + T_F(BELT)\big)$$

A function in the Metering window allows the torque indicated by the Torque meter ($T_{OPP.}$) to be corrected so that it corresponds to the opposition torque produced by the driven machine ($T_{OPP.}(MACHINE)$). This function simply subtracts the typical value of $T_F(P.M.)$ and $T_F(BELT)$ from $T_{OPP.}$ to obtain $T_{OPP.}(MACHINE)$. The corrected torque ($T_{OPP.}(MACHINE)$) is always smaller than the original uncorrected torque ($T_{OPP.}$).

Brake operation

The brake in the Prime Mover/Dynamometer module is a device which can be coupled to a drive motor using a timing belt. It is used to mechanically load the motor. Furthermore, the brake also measures the speed and output torque of the motor. In other words, it acts as a brake as well as a speed meter and dynamometer, all at the same time. The mechanical load to the motor can be changed using a torque control (MANUAL CONTROL knob) on the brake. This allows study of the behaviour of a motor under various load conditions.

As for the prime mover, the display on the Prime Mover/Dynamometer module allows direct reading of the speed and torque values. Connecting the SPEED and TORQUE OUTPUTS of the Prime Mover/Dynamometer to the corresponding inputs on the data acquisition module permits measurement and display of speed and torque using the Speed and Torque meters in the Metering window.

The displayed speed, either on the module display or the Speed meter in the Metering window, is the actual speed at which the brake rotates. It is positive for clockwise rotation and negative for counterclockwise rotation.

The mechanical load which the brake produces when it is coupled to a drive motor consists of friction in the coupling device (timing belt), friction in the brake (bearing, brushes, and windage frictions), and a magnetic torque which the brake produces to oppose rotation of the motor ($T_M(BRK.)$). The combined effect of these frictions and this torque results in a load torque (T_{LOAD}) that opposes rotation of the motor coupled to the brake, as indicated in the following equation:

$$T_{LOAD} = T_F(BELT) + T_F(BRK.) + T_M(BRK.)$$

where $T_F(BELT)$ is the torque that opposes rotation which results from the belt friction.

$T_F(BRK.)$ is the torque that opposes rotation which results from friction in the brake.

$T_M(BRK.)$ is the magnetic torque produced in the brake to oppose rotation.

Figure 1-15 illustrates the forces that oppose rotation when a drive motor is coupled to the brake.

Torques $T_F(BELT)$ and $T_F(BRK.)$, which result from friction, vary as speed varies, as is the case with torque $T_F(P.M.)$ seen in the previous subsection. On the other hand, torque $T_M(BRK.)$ does not vary with speed, but it can be varied over a range of 0 to 3 N·m (0 to 27 lbf·in) using the torque control of the brake. Therefore, the load torque T_{LOAD} can be varied by varying torque $T_M(BRK.)$.

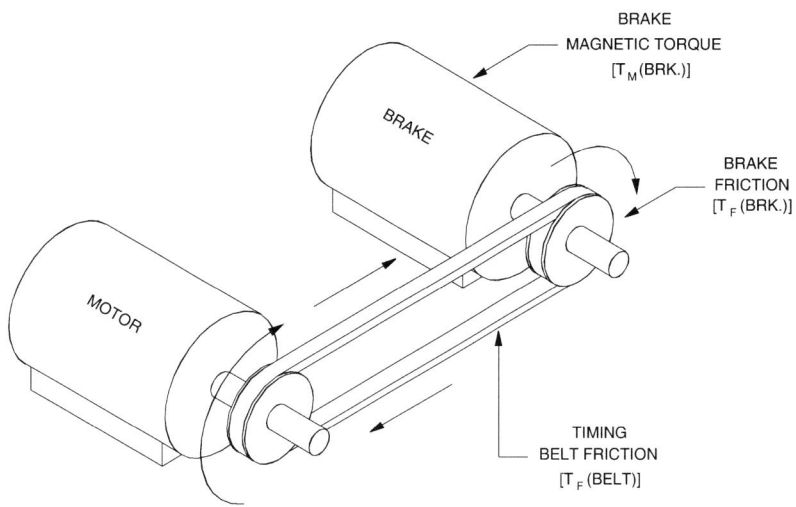

Figure 1-15. Forces that oppose motor rotation.

The torque indicated by the module display and the Torque meter (i.e., the magnetic torque) in the Metering window has the same magnitude as the magnetic torque $T_M(BRK.)$ but is of opposite polarity. In other words, the indicated torque is equal to $-T_M(BRK.)$. This means that the torque display of the Prime Mover/Dynamometer and the Torque meter indicate the torque which the motor coupled to the brake must provide to overcome the opposition to rotation caused by torque $T_M(BRK.)$. Therefore, the displayed torque for clockwise (positive) rotation is positive. For counterclockwise (negative) rotation, the displayed torque is negative. In brief, the brake torque and speed displayed are always of the same polarity. Furthermore, when the MANUAL LOAD CONTROL knob is set to minimum [fully counterclockwise, or CCW position], torque $T_M(BRK.)$ is zero, and therefore, the torque indicated is also 0. This torque increases as the LOAD CONTROL knob is turned clockwise.

However, the torque indicated by the module display and the Torque meter (i.e., the magnetic torque) does not correspond to the output torque (T_{OUT}) of the motor coupled to the brake. To rotate, this motor must produce sufficient output torque T_{OUT} to overcome the load torque T_{LOAD}, which consists of the brake magnetic torque $T_M(BRK.)$ plus the belt friction torque $T_F(BELT)$ plus the brake friction torque $T_F(BRK.)$. In other words, the motor output torque T_{OUT} must be equal to the load torque T_{LOAD} but of opposite polarity. This is indicated in the following equation:

$$T_{OUT} = -T_{LOAD} = -\left(T_M(BRK.) + T_F(BELT) + T_F(BRK.)\right)$$

A function allows the torque indicated by the Torque meter in the Metering window to be corrected so that it indicates the motor output torque T_{OUT}. This function simply adds the typical value of $T_F(BELT)$ and $T_F(BRK.)$ to $T_M(BRK.)$. The corrected torque, i.e., the motor output torque T_{OUT}, is thus always greater than the original uncorrected torque.

Procedure summary

In the first part of this exercise, you will experiment with the Prime Mover/Dynamometer operating in the prime mover mode. You will set up the equipment in the Workstation, connect the equipment as shown in Figure 1-16, and make the appropriate settings on the Prime Mover/Dynamometer. You will check that the direction of rotation of the prime mover changes when the polarity of the input voltage is changed. You will observe the effect of the MODE switch on the prime mover operation. You will then measure the prime mover friction torque. Finally, you will couple the prime mover to a squirrel-cage induction motor and measure the opposition torque caused by the motor.

In the second part of this exercise, you will experiment with the Prime Mover/Dynamometer operating in the brake mode. You will couple the brake to a dc motor. You will vary the setting of the torque command (LOAD CONTROL knob) of the brake while observing the speed and torque polarity indicated by the Speed and Torque meters in the Metering window. You will learn how to correct the torque indicated by the Torque meter to determine the output torque of the dc motor.

EQUIPMENT REQUIRED

Refer to the Equipment Utilization Chart in Appendix C to obtain the list of equipment required for this exercise.

PROCEDURE

 High voltages are present in this laboratory exercise. Do not make or modify any banana jack connections with the power on unless otherwise specified.

Setting up the equipment

1. Install the equipment in the EMS Workstation.

2. On the Power Supply, make sure the main power switch is set to the O (off) position, and the voltage control knob is turned fully counterclockwise. Ensure the Power Supply is connected to a three-phase power source.

3. Ensure that the data acquisition module is connected to a USB port of the computer.

   Connect the LOW POWER INPUTs of the data acquisition module and Prime Mover/Dynamometer to the 24 V - AC output of the Power Supply.

   On the Power Supply, set the 24 V - AC power switch to the I (on) position.

4. Start the Data Acquisition software (LVDAC or LVDAM). Open setup configuration file DCMOTOR1.DAI.

 If you are using LVSIM-EMS in LVVL, you must use the IMPORT option in the File menu to open the configuration file.

In the Metering window, select layout 1 via the View or Options menu. Make sure that the continuous refresh mode is selected.

 *If you are using LVDAC-EMS, set parameter **Analog Input AI7** in the Data Acquisition and Control Settings window to **Non-Corrected Torque (N·m)**.*

5. Connect the equipment as shown in Figure 1-16.

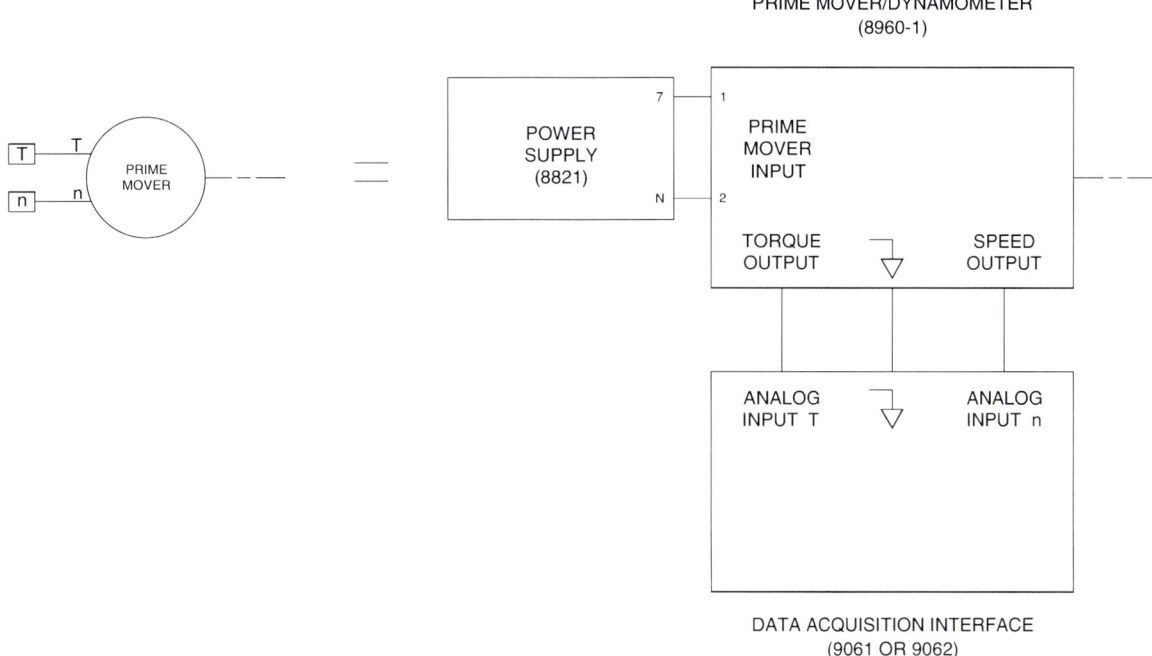

Figure 1-16. Prime mover circuit.

6. Set the Prime Mover/Dynamometer controls as follows:

   MODE switch............................PRIME MOVER (P.M.)
   DISPLAY switch...SPEED (N)

 If you are performing the exercise using LVSIM®-EMS, you can zoom in the Prime Mover/Dynamometer module before setting the controls in order to see additional front panel markings related to these controls.

These settings make the Prime Mover/Dynamometer operate as a prime mover.

Speed, polarity, and direction of rotation

7. Turn the Power Supply on by setting its main power switch to the I (ON) position, and set the voltage control knob at about 10%.

   Record the speed n indicated by the Prime Mover/Dynamometer display in the following blank space.

   n (prime mover) = _____ r/min

   Notice that the Speed meter in the Metering window also indicates the prime mover speed.

   Observe the Prime Mover/Dynamometer. What is the direction of rotation?

   Direction of rotation: _____

8. Turn the Power Supply off by setting its main power switch to the O (off) position. Do not change the setting of the voltage control knob.

   Reverse the connection of the leads at the PRIME MOVER INPUT.

   Turn the Power Supply on and observe the Prime Mover/Dynamometer. What is the direction of rotation?

   Direction of rotation: _____

   With this reversed polarity connection, what difference is there in the speed indicated by the Speed meter in the Metering window?

   _____

   _____

In this manual, clockwise rotation is defined as the positive direction, and counterclockwise rotation is therefore indicated by a negative speed value. The speed indicated by the Prime Mover/Dynamometer display and the Speed meter in the Metering window uses this same convention.

Effect of the MODE switch

9. Turn the voltage control knob fully counterclockwise and turn the Power Supply off.

   Reconnect the leads at the PRIME MOVER INPUT as they were in step 7.

10. Turn the Power Supply on and set the voltage control knob at about 10%.

    On the Prime Mover/Dynamometer, set the MODE switch to the DYN. position then wait a few seconds.

    Does the prime mover stop rotating, thus showing that power has been cut off?

    ❏ Yes   ❏ No

Measuring the prime mover friction torque

11. On the Prime Mover/Dynamometer, set the MODE switch back to the PRIME MOVER (P.M.) position.

    On the Power Supply, set the voltage control knob so that the prime mover rotates at a speed of 1500 r/min.

12. On the Prime Mover/Dynamometer, set the DISPLAY switch to the TORQUE (T) position.

    Record the friction torque ($T_F(P.M.)$) indicated by the prime mover display.

    $T_F(P.M.) = $ _____ N · m (lbf · in) $[n = 1500 \text{ r/min}]$

    Why is the torque indicated by the Prime Mover/Dynamometer display negative while the prime mover speed is positive (clockwise rotation)?

    _____

    _____

    Notice that the Torque meter in the Metering window indicates approximately the same torque as the module display.

    On the Prime Mover/Dynamometer, set the DISPLAY switch back to the SPEED (N) position.

Ex. 1-2 – Prime Mover and Brake Operation (Model 8960-1) ♦ *Procedure*

13. On the Power Supply, turn the voltage control knob fully counterclockwise and turn the Power Supply off.

    Reverse the connection of the leads at the PRIME MOVER INPUT.

    Turn the Power Supply on and set the voltage control knob so that the prime mover rotates at a speed of -1500 r/min.

    Is the torque displayed on the Torque meter in the Metering window ($T_F(P.M.)$) of opposite sign but approximately equal to the value recorded for the same speed?

    ❏ Yes   ❏ No

Measuring the opposition torque caused by the driven machine

14. On the Power Supply, turn the voltage control knob fully counterclockwise and turn the Power Supply off.

    Reverse the connection of the leads at the PRIME MOVER INPUT.

    Use a timing belt to mechanically couple the prime mover/dynamometer module to the Four-Pole Squirrel-Cage Induction Motor.

    ⚠ WARNING

    Before installing or removing a timing belt, make absolutely sure that power is turned off to prevent any rotating machine from starting inadvertently.

15. Turn the Power Supply on and set the voltage control knob so that the prime mover rotates at a speed of 1500 r/min.

    Record the opposition torque ($T_{OPP.}$) indicated by the Torque meter in the Metering window.

    $T_{OPP.} =$ _____ N · m (lbf · in)   [$n = 1500$ r/min]

    Compare the opposition torque measured in this step to the opposition torque ($T_F(P.M.)$) recorded in step 12. Briefly explain the difference between these two opposition torques.

    _____
    _____
    _____

16. In the Metering window, enable the torque correction function for the Torque meter by clicking the button in the lower left corner of the meter. The indication "C" appears in this button to indicate that the torque correction is enabled. The Torque meter now indicates the opposition torque caused by the Four-Pole Squirrel-Cage Induction Motor $T_{OPP.}(MACHINE)$. Record this torque in the following blank space.

$T_{OPP.}(MACHINE) =$ _____ N · m (lbf · in)   [$n = 1500$ r/min]

Using the torque measured in this step and the previous step, compare the opposition torque caused by the Four-Pole Squirrel-Cage Induction Motor ($T_{OPP.}(MACHINE)$) to the total opposition torque ($T_{OPP.}$).

_____

_____

_____

17. Turn the Power Supply off.

If you are performing the exercise using LVSIM-EMS, go to the next subsection of this exercise. Otherwise continue this step.

On the Four-Pole Squirrel-Cage Induction Motor, tilt the front panel forward to get access to the motor's shaft.

Turn the motor's shaft manually. While doing this, notice the twisting force you must apply to make the motor's shaft rotate.

⚠ WARNING

Before turning a motor's shaft manually or removing a timing belt, make absolutely sure that power is turned off to prevent any rotating machine from starting inadvertently.

Remove the timing belt that couples the Prime Mover/Dynamometer to the Four-Pole Squirrel-Cage Induction Motor.

Turn the motor's shaft manually. While doing this, notice the twisting force you must apply to make the motor's shaft rotate. Is this force much smaller than that required to make the motor's shaft rotate when the motor is coupled to the Prime Mover/Dynamometer module?

❑ Yes   ❑ No

Is your observation the same as when you compared the opposition torques in step 16? Briefly explain.

_____

_____

_____

Brake operation

18. Remove the Four-Pole Squirrel-Cage Induction Motor from the EMS Workstation. Install a DC Motor/Generator in the EMS workstation (next to the Prime Mover/Dynamometer).

Use a timing belt to mechanically couple the Prime Mover/Dynamometer to the DC Motor/Generator.

Connect the equipment as shown in Figure 1-17.

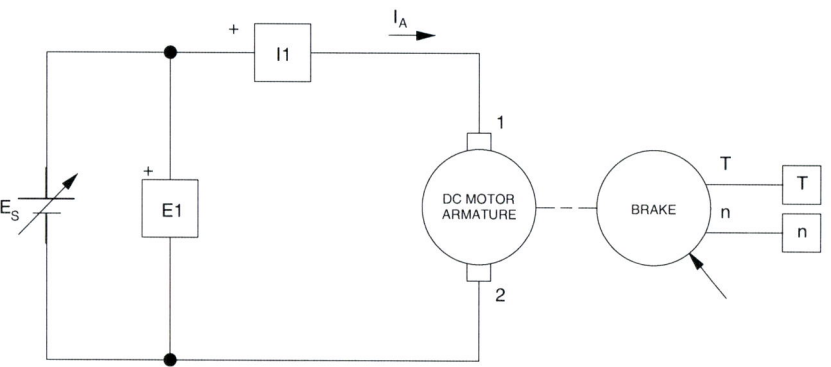

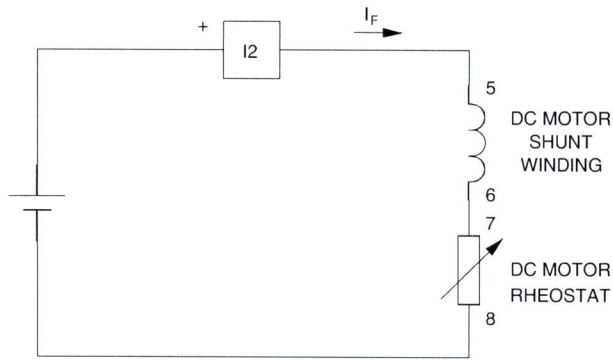

Figure 1-17. DC motor coupled to a brake.

19. On the Prime Mover/Dynamometer module, set the controls as follows:

    MODE switch..DYN.
    LOAD CONTROL MODE switch......................MAN.
    LOAD CONTROL knob...................MIN. (fully CCW)
    DISPLAY switch.....................................TORQUE (T)

These settings make the Prime Mover/Dynamometer operate as a brake.

 If you are performing the exercise using LVSIM®-EMS, you can zoom in the Prime Mover/Dynamometer module before setting the controls in order to see additional front panel markings related to these controls.

20. Turn the Power Supply on.

On the DC Motor/Generator, set the FIELD RHEOSTAT so that the current indicated by meter I field (I_f) in the Metering window is equal to the value given in Table 1-2.

Table 1-2. DC motor field current.

Local ac power network		I_f
Voltage (V)	Frequency (Hz)	(mA)
120	60	300
220	50	190
220	60	190
240	50	210

On the Power Supply, set the voltage control knob so that the dc motor rotates at a speed of 1500 r/min. The display of the Prime Mover/Dynamometer and the Speed meter in the Metering window indicate the motor speed.

In the Metering window, disable the torque correction function for the Torque meter by clicking the button in the lower left corner of the meter. The indication "NC" appears in this button when the torque correction is disabled.

Observe that the torque indicated by the display of the Prime Mover/Dynamometer and the Torque meter in the Metering window is zero. Briefly explain why.

_____

_____

21. On the Prime Mover/Dynamometer, slowly turn the LOAD CONTROL knob (torque control) clockwise until the torque indicated by the module display and the Torque meter in the Metering window is equal to 1.0 N·m (9.0 lbf·in). While doing this, observe the speed indicated by the Speed meter in the Metering window.

What happens to the speed as the torque passes from 0 to 1.0 N·m (0 to 9.0 lbf·in)?

_____

_____

Ex. 1-2 – Prime Mover and Brake Operation (Model 8960-1) ♦ *Procedure*

Notice that the torque has the same polarity as the speed n. Why?

_____
_____
_____

Measuring the output torque of a drive motor

22. On the Power Supply, readjust the voltage control knob so that the dc motor rotates at a speed of 1500 r/min.

    Make sure that the torque indicated by the display of the Prime Mover/Dynamometer and the Torque meter in the Metering window is equal to 1.0 N·m (9.0 lbf·in). Slightly readjust the torque control if necessary.

    Record the dc motor torque indicated by the Torque meter in the following blank space.

    T_{OUT} (uncorrected) = _____ N · m (lbf · in)

    $[n = 1500 \text{ r/min}]$

    $[T_M(BRK.) = 1 \text{ N} \cdot \text{m } (9.0 \text{ lbf} \cdot \text{in})]$

23. In the Metering window, select the torque correction function for the Torque meter by clicking the button in the lower left corner of the meter. The indication "C" appears in this button to indicate that the torque correction is enabled. The Torque meter now indicates the dc motor output torque T_{OUT} (corrected). Record this torque in the following blank space.

    T_{OUT}(corrected) = _____ N · m (lbf · in)

    $[n = 1500 \text{ r/min}]$

    $[T_M(BRK.) = 1 \text{ N} \cdot \text{m } (9.0 \text{ lbf} \cdot \text{in})]$

    What happened to the torque indicated by the Torque meter when you selected the torque correction function? Briefly explain why.

    _____
    _____
    _____

24. On the Power Supply, turn the voltage control knob fully counterclockwise and set the main power switch to the O (off) position.

    Reverse the connection of the leads at the armature (terminals 1 and 2) of the DC Motor/Generator.

    Turn the Power Supply on and set the voltage control knob so that the dc motor rotates at a speed of -1500 r/min.

    Record the dc motor output torque T_{OUT} (corrected) in the following blank space.

    T_{OUT} (corrected) = _____ N · m (lbf · in)

    $[n = 1500 \text{ r/min}]$

    $[T_M(BRK.) = 1 \text{ N} \cdot \text{m } (9.0 \text{ lbf} \cdot \text{in})]$

    What effect does changing the direction of rotation have on the dc motor output torque T_{OUT} (corrected)?

    _____

    _____

25. On the Power Supply, set the main power switch and the 24 V - AC power switch to the O (off) position. Remove all leads and cables.

CONCLUSION

In the first part of this exercise, you learned that the prime mover in the Prime Mover/Dynamometer operates basically like a linear voltage-to-speed converter. The prime mover speed and direction of rotation can be changed by changing the magnitude and polarity of the input voltage, respectively. You saw that the speed indicated on the Speed meter in the Metering window is positive for clockwise rotation, and negative for counterclockwise rotation. The torque indicated on the Torque meter in the Metering window is the torque that opposes the prime mover rotation (opposition torque). A torque correction function can be enabled so that the meter indicates the opposition torque caused by the driven machine.

In the second part of this exercise, you learned that the brake in the Prime Mover/Dynamometer is used to mechanically load a motor. Furthermore, the brake also measures the speed and output torque of the motor. In other words, it acts as a brake as well as a speed meter and dynamometer, all at the same time. You saw that the mechanical load (torque command) to the motor can be changed by using the LOAD CONTROL knob on the Prime Mover/Dynamometer. When the torque command is increased, the motor speed decreases due to the increased mechanical load on the motor. The Torque meter in the Metering window indicates the torque which the motor must produce to overcome the opposition torque produced by the brake. A torque correction function can be enabled so that the meter indicates the output torque of the motor.

Ex. 1-2 – Prime Mover and Brake Operation (Model 8960-1) ◆ *Review Questions*

REVIEW QUESTIONS

1. What does a negative speed value on the prime mover indicate?

    a. That rotation is in the clockwise direction.
    b. That rotation is in the counterclockwise direction.
    c. That the prime mover is supplying torque to drive a load.
    d. That the prime mover is receiving power from a load.

2. Why does the prime mover display show a torque value even with no mechanical load applied to the machine's shaft?

    a. The display is probably defective.
    b. Because the prime mover must supply torque to overcome friction.
    c. This indicates the electrical power supplied to the prime mover.
    d. This indicates the mechanical power supplied to the prime mover.

3. Will the torque indicated by the prime mover display be a negative or positive value for clockwise rotation?

    a. Negative.
    b. Positive.
    c. It depends on how fast the prime mover rotates.
    d. It depends on the value of the applied voltage.

4. In the brake mode, what does the display of the Prime Mover/Dynamometer indicate when the DISPLAY switch is set to the TORQUE (T) position?

    a. The sum of the brake friction torque $T_F(BRK.)$ and belt friction torque $T_F(BELT)$.
    b. The magnetic torque produced by the brake to oppose rotation ($T_M(BRK.)$).
    c. The torque which the motor coupled to the brake must produce to overcome the magnetic torque $T_M(BRK.)$, i.e., the inverse of $T_M(BRK.)$.
    d. The load torque T_{LOAD} produced by the brake.

5. The Torque meter in the Metering window is used to display torque when the brake is coupled to a drive motor. The torque correction function is enabled. What does the Torque meter now indicate?

    a. The load torque T_{LOAD}.
    b. The sum of the belt friction torque $T_F(BELT)$, brake friction torque $T_F(BRK.)$, and brake magnetic torque $T_M(BRK.)$.
    c. The output torque T_{OUT} of the motor coupled to the brake minus the belt friction torque $T_F(BELT)$.
    d. The output torque T_{OUT} of the motor coupled by the brake.

Exercise 1-3

Motor Power, Losses, and Efficiency

EXERCISE OBJECTIVE   When you have completed this exercise, you will be able to determine motor power, losses, and efficiency using a brake.

DISCUSSION   Torque was earlier defined as a twisting force that causes an object to rotate. In electric motors, this twisting force comes from the interaction of magnetic fields, and its value is related to the current flowing in the motor. Since the magnetic forces in the rotor of a dc motor are produced by current flowing in a wire loop, increasing the current will increase the strength of the magnetic forces. The motor will therefore produce more torque, meaning increased **motor power**, and it will consume more electric power.

The prime mover that you have been using is actually a dc motor and it converts electrical power to mechanical power. Figure 1-18 gives an overview of the power flow and power losses in a dc motor.

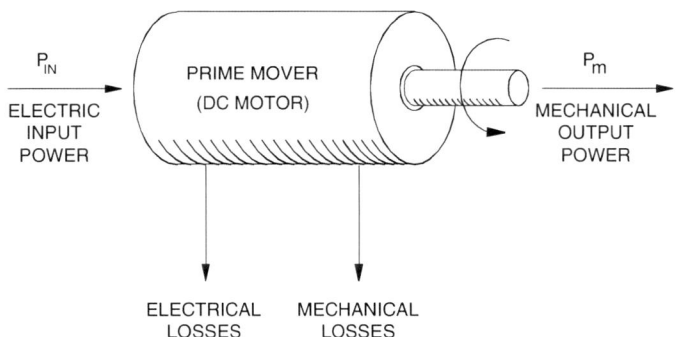

Figure 1-18. Typical power flow in a dc motor.

Electric **motor efficiency** is expressed as the ratio of its mechanical output power to its electrical input power, P_m/P_{in}. The mechanical output power of a motor depends on its speed and torque, and can be determined using one of the following two formulas, depending on whether torque is expressed in N·m or lbf·in:

$$P_m = \frac{n \times T}{9.55} \quad \text{(torque expressed in N} \cdot \text{m)}$$

$$P_m = \frac{n \times T}{84.51} \quad \text{(torque expressed in lbf} \cdot \text{in)}$$

Ex. 1-3 – Motor Power, Losses, and Efficiency ♦ *Discussion*

Efficiency for a motor is usually shown in the form of a graph of efficiency versus mechanical output power, although a specific value at the nominal power rating is sometimes given.

Rotating machine losses fall into two categories, mechanical losses and electrical losses. Mechanical losses come from bearing friction, brushes friction, as well as windage or cooling-fan friction. These losses vary somewhat as speed increases from zero to its nominal value but remain fairly constant over the normal operating range between no-load and full-load. A typical losses and efficiency versus mechanical output power graph for a 10-kW dc motor is shown in Figure 1-19.

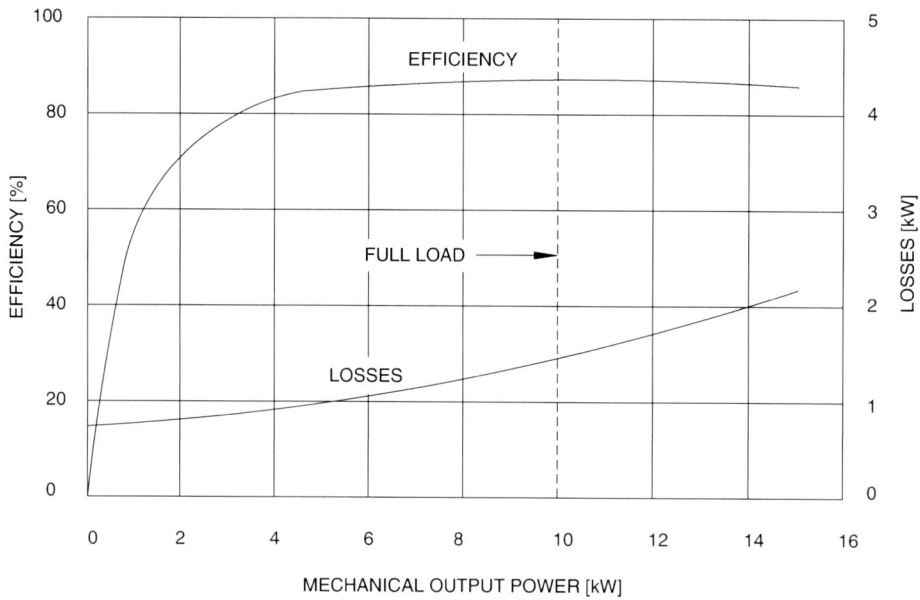

Figure 1-19. **Typical losses and efficiency graph for a 10-kW dc motor.**

Electrical losses are classed as copper losses, brushes losses, and iron losses. Copper losses (I^2R) result from the resistance of the wire used in the machine, are dissipated as heat, and depend on the value of current in the machine. Brushes losses are usually very small, and are due to the contact resistance of the brush which causes a typical voltage drop between 0.8 V and 1.3 V. Finally, iron losses come from hysteresis and eddy currents in the machine, and depend on the magnetic flux density, the speed of rotation or frequency, the kind of steel and the size of the motor.

Procedure summary

In the first part of the exercise, you will set up the equipment in the Workstation, connect the equipment as shown in Figure 1-20, and make the appropriate settings on the equipment.

Ex. 1-3 – Motor Power, Losses, and Efficiency ♦ Procedure

In the second part of the exercise, you will set the dc motor speed to 1500 r/min and adjust the torque setting of the brake to set the mechanical load applied to the dc motor. You will then measure the dc motor speed, output torque (T_{OUT}), and electrical input power (P_{in}). You will use these results to calculate the dc motor mechanical output power (P_m) and efficiency (η), as well as the amount of power lost in the dc motor. You will compare the calculated mechanical output power and efficiency to those measured with meters in the Metering window.

In the third part of the exercise, you will vary the mechanical load applied to the dc motor in 0.2-N·m (1.5 lbf·in) steps. For each step, you will record data in the Data Table to plot a graph of the dc motor efficiency versus the mechanical output power.

EQUIPMENT REQUIRED

Refer to the Equipment Utilization Chart in Appendix C to obtain the list of equipment required for this exercise.

PROCEDURE

High voltages are present in this laboratory exercise. Do not make or modify any banana jack connections with the power on unless otherwise specified.

Setting up the equipment

1. Install the equipment required in the EMS workstation, making sure that the DC Motor/Generator is installed to the right of the prime mover/dynamometer module (Four-Quadrant Dynamometer/Power Supply, Model 8960-2, or Prime Mover/Dynamometer, Model 8960-1).

    If you are performing the exercise using the EMS system, ensure that the brushes of the DC Motor/Generator are adjusted to the neutral point. To do so, connect an ac power source (terminals 4 and N of the Power Supply) to the armature of the AC Motor/Generator (terminals 1 and 2) through CURRENT INPUT I1 of the data acquisition module. Connect the shunt winding of the DC Motor/Generator (terminals 5 and 6) to VOLTAGE INPUT E1 of the data acquisition module. Start the Metering application and open setup configuration file ACMOTOR1.DAI. Turn the Power Supply on and set the voltage control knob so that an ac current (indicated by meter I line 1) equal to half the nominal value of the armature current flows in the armature of the DC Motor/Generator. Adjust the brush adjustment lever on the DC Motor/Generator so that the voltage across the shunt winding (indicated by meter E line 1) is minimum. Turn the Power Supply off, exit the Metering application, and disconnect all leads and cable.

   Mechanically couple the prime mover/dynamometer module to the DC Motor/Generator using a timing belt.

2. On the Power Supply, make sure the main power switch is set to the O (off) position, and the voltage control knob is turned fully counterclockwise. Ensure the Power Supply is connected to a three-phase power source.

 If you are using the Four-Quadrant Dynamometer/Power Supply, Model 8960-2, connect its POWER INPUT to a wall receptacle.

3. Ensure that the data acquisition module is connected to a USB port of the computer.

   Connect the POWER INPUT of the data acquisition module to the 24 V - AC output of the Power Supply.

 If you are using the Prime Mover/Dynamometer, Model 8960-1, connect its LOW POWER INPUT to the 24 V - AC output of the Power Supply.

   On the Power Supply, set the 24 V - AC power switch to the I (on) position.

 If you are using the Four-Quadrant Dynamometer/Power Supply, Model 8960-2, turn it on by setting its POWER INPUT switch to the I (on) position. Press and hold the FUNCTION button 3 seconds to have uncorrected torque values on the display of the Four-Quadrant Dynamometer/Power Supply. The indication "NC" appears next to the function name on the display to indicate that the torque values are uncorrected.

4. Start the Data Acquisition software (LVDAC or LVDAM). Open setup configuration file DCMOTOR1.DAI.

 If you are using LVSIM-EMS in LVVL, you must use the IMPORT option in the File menu to open the configuration file.

   In the Metering window, select layout 3. Make sure that the continuous refresh mode is selected.

5. Connect the modules as shown in Figure 1-20.

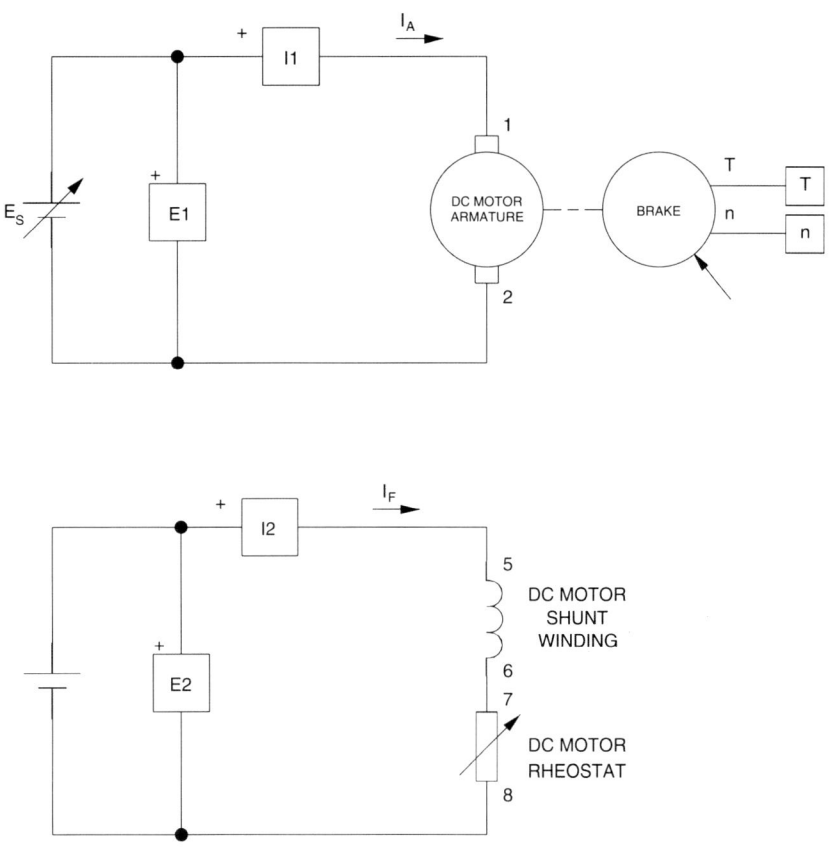

Figure 1-20. Setup for measuring the dc motor power, losses, and efficiency.

6. Turn the Power Supply on.

On the DC Motor/Generator, set the FIELD RHEOSTAT so that the field current I_f indicated in the Metering window is equal to the value given in Table 1-3. The field current I_f flows in the shunt winding of the dc motor and is necessary for its operation. This will be explained later in this manual.

Ex. 1-3 – Motor Power, Losses, and Efficiency ♦ *Procedure*

Table 1-3. DC motor field current I_f.

Local ac power network		I_f (mA)
Voltage (V)	Frequency (Hz)	
120	60	300
220	50	190
220	60	190
240	50	210

Set the Four-Quadrant Dynamometer/Power Supply or the Prime Mover/Dynamometer to operate as a brake, then set the brake torque control to minimum (knob fully counterclockwise position). If necessary, refer to Exercise 1-1 or Exercise 1-2 (depending on whether you are using Model 8960-2 or Model 8960-1) to do this.

 If you are performing the exercise using LVSIM®-EMS, you can zoom in on the Prime Mover/Dynamometer before setting the controls in order to see additional front panel markings related to these controls.

In the Metering window, make sure that the **torque correction** function of the Torque meter **is enabled**.

Prime mover efficiency measurement

7. On the Power Supply, set the voltage control knob so that the dc motor rotates at a speed of 1500 r/min.

   On the brake, set the torque control so that the torque indicated by the Torque meter in the Metering window is equal to 1.3 N·m (11.7 lbf·in).

8. Record the dc motor speed (n) and output torque (T_{OUT}) in the following blank spaces. These are indicated by the Speed and Torque meters in the Metering window.

   n = _____ r/min

   T_{OUT} = _____ N·m (lbf·in)

Ex. 1-3 – Motor Power, Losses, and Efficiency ♦ *Procedure*

Calculate the dc motor mechanical output power (P_m), using the measured speed n and output torque T_{OUT}, and one of the following formulas, depending on whether torque is expressed in N·m or lbf·in:

$$P_m = \frac{n \times T_{OUT}}{9.55} \quad \text{(torque expressed in N} \cdot \text{m)}$$

$$P_m = \frac{n \times T_{OUT}}{84.51} \quad \text{(torque expressed in lbf} \cdot \text{in)}$$

$P_m = $ _____ W

Record in the following blank space the dc motor mechanical output power P_m indicated by the Mech. Power meter in the Metering window.

$P_m = $ _____ W (measured)

Compare the calculated and measured values of the dc motor mechanical output power. Are they approximately equal?

❏ Yes   ❏ No

9. Record in the following blank space the electrical power supplied to the dc motor armature (P_A). It is indicated by meter P arm. (P_A) in the Metering window.

$P_A = $ _____ W

Record in the following blank space the dc motor field power P_{field}. It is indicated by meter P_{field} in the Metering window.

$P_{field} = $ _____ W

Calculate the total amount of power (P_{in}) supplied to the dc motor ($P_A + P_{field}$). Record your result below.

$P_{in} = P_A + P_{field} = $ _____ W

Compare the dc motor mechanical output power P_m (measured in step 8) to the electrical input power P_{in} (calculated above). How much power is lost in the dc motor?

_____

_____

10. Calculate the dc motor efficiency (η) using the dc motor electrical input power P_{in} and mechanical output power P_m, and the following formula:

$$\eta = \frac{P_m}{P_{in}} \times 100\%$$

$\eta =$ _____ % (calculated)

Record in the following blank space the dc motor efficiency η indicated by meter $P_m / (P_{in} + P_{field})$ in the Metering window.

$\eta =$ _____ % (measured)

Compare the calculated and measured values of the dc motor efficiency η. Are they approximately equal?

❏ Yes   ❏ No

The dc motor efficiency may seem to be rather low, but it is typical of efficiency values obtained with small motors. Generally, motors having a nominal power of less than 10 kW have efficiencies ranging from about 60% to 85%. For motors with nominal power above 10 kW, efficiencies as high as 98% can be obtained.

Efficiency versus mechanical output power

11. On the brake, set the torque control to minimum (knob turned fully counterclockwise). The torque indicated by the module display should be equal to 0.0 N·m (0.0 lbf·in).

On the Power Supply, slightly readjust the voltage control knob so that the dc motor speed is equal to 1500 r/min (if necessary).

12. Record the dc motor voltage, dc motor current, electrical input power, field voltage, field power, output torque, speed, mechanical output power, and efficiency in the Data Table. These parameters are indicated by meters E arm. (E_A), I arm. (I_A), P_{in}, E_{field}, P_{field}, Torque (T), Speed (n), Mech. power, and $P_m/(P_{in} + P_{field})$, respectively.

    On the brake, set the torque control so that the torque indicated by the Torque meter in the Metering window increases by 0.2 N·m (1.5 lbf·in) increments up to 2.0 N·m (18.0 lbf·in). For each torque setting, record the dc motor voltage, current, electrical input power, field voltage, field power, output torque, speed, mechanical output power, and efficiency in the Data Table.

13. When all data has been recorded, turn the voltage control knob fully counterclockwise and turn the Power Supply off.

    In the Data Table window, confirm that the data has been stored, entitle the Data Table as DT131, and print the Data Table.

    Refer to the user guide dealing with the computer-based instruments for EMS to know how to entitle and print a Data Table.

14. In the Graph window, make the appropriate settings to obtain a graph of the dc motor efficiency [obtained from meter $P_m/(P_{in} + P_{field})$] as a function of the dc motor mechanical output power (obtained from meter Mech. Power). Entitle the graph as G131, name the x-axis as Prime mover (dc motor) mechanical output power, name the y-axis as Prime mover (dc motor) efficiency, and print the graph.

    Refer to the user guide dealing with the computer-based instruments for EMS to know how to use the Graph window of the Metering application to obtain a graph, entitle a graph, name the axes of a graph, and print a graph.

    Describe how the dc motor efficiency varies as a function of the mechanical output power.

    _____

    _____

    _____

    Compare graph G131 with the graph shown in Figure 1-19.

    _____

    _____

15. On the Power Supply, set the 24 V - AC power switch to the O (off) position.

 If you are using the Four-Quadrant Dynamometer/Power Supply, Model 8960-2, turn it off by setting its POWER INPUT switch to the O (off) position.

Remove all leads and cables.

CONCLUSION

This exercise allowed you to calculate the mechanical output power produced by a motor using its speed and output torque. You determined motor efficiency by calculating the ratio of mechanical output power to electrical input power, and produced an efficiency versus mechanical output power graph.

REVIEW QUESTIONS

1. What is the formula for calculating the mechanical output power of a motor?

    a. $P_m = (n \times T)/9.55$ (torque in N·m)

    $P_m = (n \times T)/84.51$ (torque in lbf·in)

    b. $P_m = n \times T \times 9.55$ (torque in N·m)

    $P_m = n \times T \times 84.51$ (torque in lbf·in)

    c. $P_m = n/(9.55 \times T)$ (torque in N·m)

    $P_m = n/(84.51 \times T)$ (torque in lbf·in)

    d. $P_m = (n/T) \times 9.55$ (torque in N·m)

    $P_m = (n/T) \times 84.51$ (torque in lbf·in)

2. What is the definition of motor efficiency?

    a. It is the rate of doing work.
    b. It is the amount of work produced.
    c. It is the ratio of output mechanical power to input electrical power.
    d. It is the difference between input and output power in watts.

3. A dc motor turns at a speed of 1460 r/min and produces an output torque of 23.5 N·m (208 lbf·in). The dc voltage applied to the motor is 280 V and a current of 14.1 A flows through the motor. What is the efficiency of the motor?

    a. 94%
    b. 91%
    c. 79%
    d. 86%

Ex. 1-3 – Motor Power, Losses, and Efficiency ◆ *Review Questions*

4. Using the data of the previous question, calculate the amount of power lost in the motor.

    a. 829 W
    b. 553 W
    c. 355 W
    d. 237 W

5. What are the two main categories of losses for rotating machines?

    a. Copper losses and iron losses.
    b. Electrical losses and copper losses.
    c. Mechanical losses and iron losses.
    d. Electrical losses and mechanical losses.

Unit Test

1. When the prime mover/dynamometer module operates as a prime mover, why does the Torque meter in the Metering window indicate a torque value even with no load applied to the machine's shaft?

    a. Because the prime mover must supply torque to overcome friction.
    b. The prime mover display is probably defective.
    c. This indicates the mechanical power supplied to the prime mover.
    d. This indicates the electrical power supplied to the prime mover.

2. A moving loop of wire cuts a magnetic field. Knowing that the magnetic flux linking the loop passes from 0 to 280 mWb in 0.05 second as the loop cuts the magnetic field, what is the voltage induced across the loop of wire?

    a. 14 V
    b. 5.6 kV
    c. 5.6 V
    d. 0.014 V

3. A prime mover is mechanically coupled to a brake. What happens to the prime mover speed when the brake torque control is adjusted so as to decrease the load torque T_{LOAD}?

    a. The speed decreases slowly until the motor stops rotating.
    b. The speed does not change.
    c. The speed increases.
    d. The speed becomes unstable because the prime mover reacts strongly to load torque variations.

4. When the prime mover/dynamometer module operates as a brake and the brake is mechanically coupled to a drive motor, what does the Torque meter in the Metering window indicate when the torque correction function is disabled?

    a. The load torque T_{LOAD} produced by the brake.
    b. The magnetic torque produced by the brake to oppose rotation ($T_M\ (BRK)$).
    c. The sum of the brake friction torque $T_F\ (BRK.)$ and belt friction torque $T_F\ (BELT)$.
    d. The torque which the drive motor coupled to the brake must produce to overcome the magnetic torque $T_M\ (BRK.)$, i.e., the inverse of $T_M\ (BRK.)$.

5. When the prime mover/dynamometer module operates as a brake and the brake is mechanically coupled to a drive motor, what does the Torque meter in the Metering window indicate when the torque correction function is enabled?

   a. The output torque T_{OUT} of the drive motor coupled to the brake minus the belt friction torque T_F (BELT).
   b. The sum of the belt friction torque T_F (BELT), brake friction torque T_F (BRK.), and brake magnetic torque T_M (BRK.).
   c. The output torque T_{OUT} of the drive motor coupled to the brake.
   d. The load torque T_{LOAD}.

6. The motor of a water pump produces an output torque equal to 10 N·m (88.5 lbf·in). How much work is done by the pump motor if it rotates at a speed of 3000 r/min during 10 minutes?

   a. 31.4 kW
   b. 3.14 kJ
   c. 1.88 MJ
   d. 31.4 kJ

7. A prime mover is mechanically coupled to a brake. What happens to the prime mover speed when the torque control of the brake is increased in order to increase the load torque T_{LOAD}?

   a. The speed does not change.
   b. The speed decreases slowly until the motor stops rotating.
   c. The speed increases because the prime mover reacts strongly to load torque variations.
   d. The speed decreases.

8. A prime mover is mechanically coupled to a brake. The torque control on the brake is changed so that the torque produced by the prime mover passes from 0.4 to 2.6 N·m (3.5 to 23.0 lbf·in). By how much did the magnetic torque T_M (BRK.) produced by the brake vary?

   a. The magnetic torque T_M (BRK.) did not vary because the torque produced by the prime mover is not dependent on torque T_M (BRK.).
   b. The magnetic torque T_M (BRK.) increased by 2.2 N·m (19.5 lbf·in).
   c. The magnetic torque T_M (BRK.) decreased by 2.2 N·m (19.5 lbf·in).
   d. None of the above

9. A dc motor turns at a speed of 530 r/min and produces an output torque of 162 N·m (1434 lbf·in). The dc voltage applied to the motor is 280 V. Knowing that the power lost in the motor is 473 W, what are the motor efficiency η and the current I_M flowing in the motor?

   a. $\eta = 91\%$, $I_M = 33.8$ A
   b. $\eta = 95\%$, $I_M = 32.1$ A
   c. $\eta = 91\%$, $I_M = 32.1$ A
   d. $\eta = 95\%$, $I_M = 33.8$ A

10. Is motor power greater or smaller when it drives a load at a higher speed?

    a. It is greater.
    b. It is smaller.
    c. Neither, because power and speed are independent of each other.
    d. It depends on whether rotation is clockwise or counterclockwise.

Unit 2

DC Motors and Generators

UNIT OBJECTIVE

When you have completed this unit, you will be able to use the DC Motor/Generator module to demonstrate and explain the operation of dc motors and generators

DISCUSSION OF FUNDAMENTALS

Operating principle of dc motors

As stated in Unit 1, motors turn because of the interaction between two magnetic fields. This unit will discuss how these magnetic fields are produced in dc motors, and how magnetic fields induce voltage in dc generators.

The basic principle of a dc motor is the creation of a rotating magnet inside the mobile part of the motor, the rotor. This is accomplished by a device called the **commutator** which is found on all dc machines. The commutator produces the alternating currents necessary for the creation of the rotating magnet from dc power provided by an external source. Figure 2-1 illustrates a typical dc motor rotor with its main parts. This figure shows that the electrical contact between the segments of the commutator and the external dc source is made through brushes. Note that the rotor of a dc motor is also referred to as the armature.

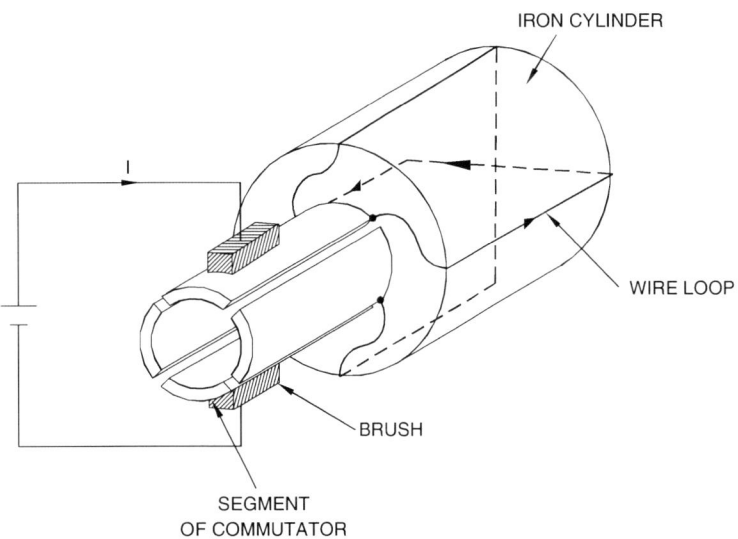

Figure 2-1. The main parts of a dc motor rotor (armature).

In Figure 2-2a, the brushes make contact with segments A and B of the commutator and current flows in wire loop A-B. No current flows in the other wire loop (C-D). This creates an electromagnet A-B with north and south poles as shown in Figure 2-2a.

If the rotor is turned clockwise a little as shown in Figure 2-2b, current still flows in wire loop A-B and the magnetic north and south poles rotate clockwise.

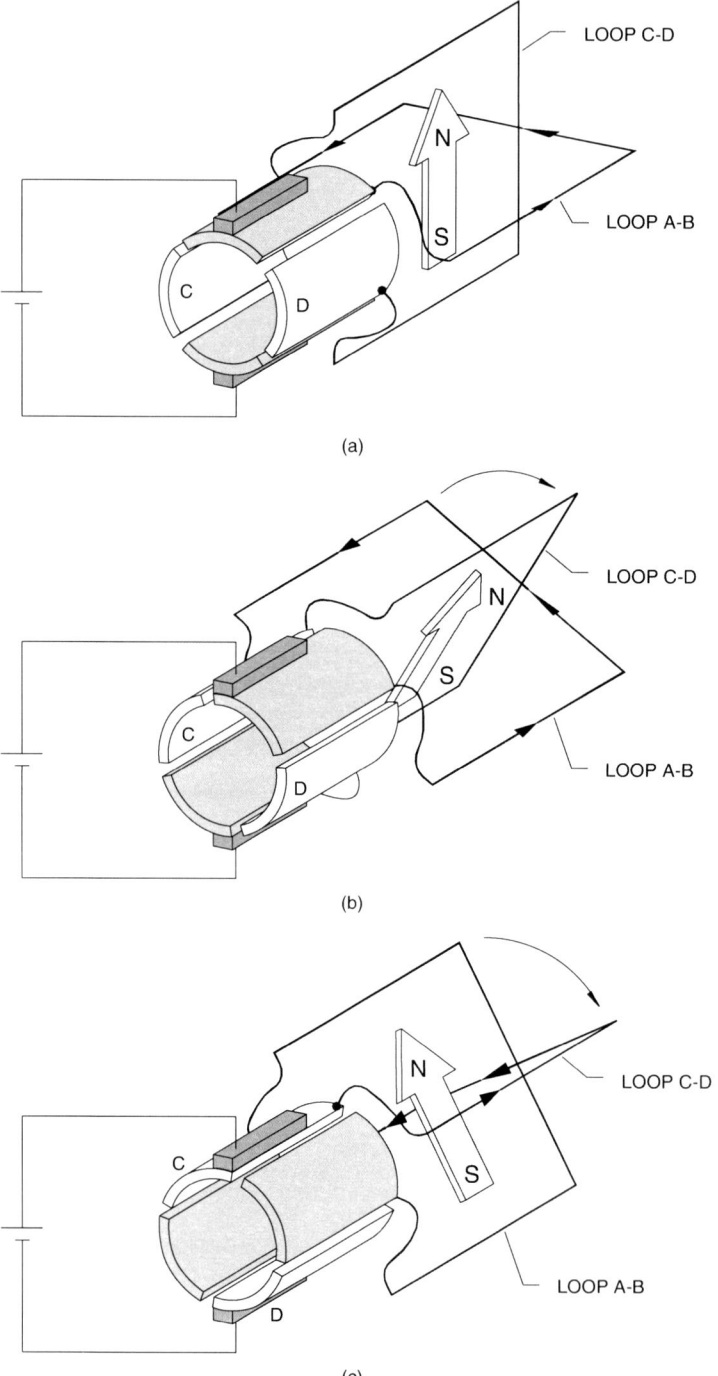

Figure 2-2. Operation of the commutator.

As the rotor continues to rotate clockwise, a time comes where a commutation occurs, i.e., the brushes make contact with segments C and D instead of segments A and B, as shown in Figure 2-2c. As a result, current now flows in wire loop C-D instead of flowing in wire loop A-B. This creates an electromagnet C-D with north and south poles as shown in Figure 2-2c.

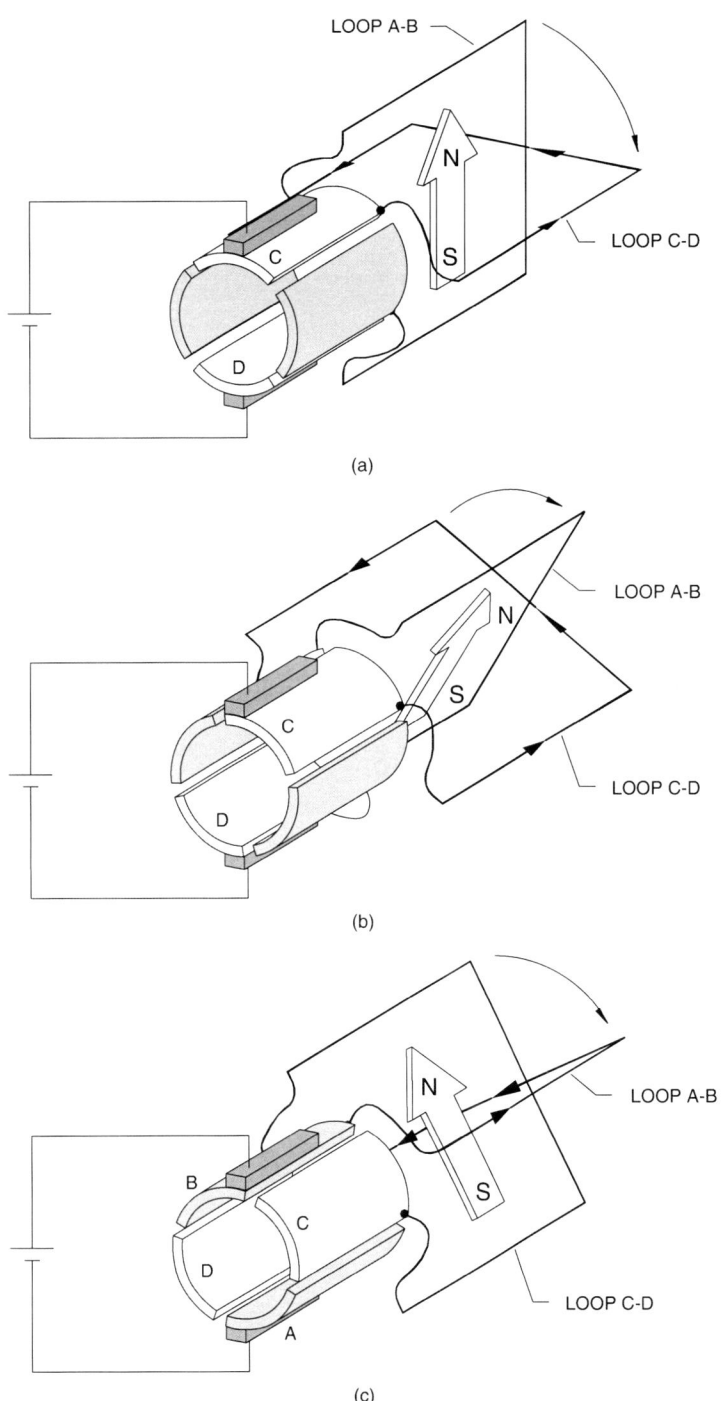

Figure 2-3. Operation of the commutator (continued).

By comparing Figure 2-2b and Figure 2-2c, you can see that the magnetic north and south poles rotate 90° counterclockwise at the commutation. As the rotor continues to rotate clockwise, the same phenomenon repeats every 90° angle of rotation, as shown in Figure 2-3a, Figure 2-3b, and Figure 2-3c.

In brief, as the rotor turns, the north and south poles of the electromagnet go back and forth (oscillate) over a 90° angle, as shown in Figure 2-4. In other words, the north and south poles are stationary, i.e., they do not rotate as the rotor turns. This is equivalent to having an electromagnet in the rotor that rotates at the same speed as the rotor but in the opposite direction. The higher the number of segments on the commutator, the lower the angle of rotation between each commutation, and the lower the angle over which the north and south poles oscillate. For example, the north and south poles would oscillate over an angle of only 11.25° if the commutator in Figure 2-2, Figure 2-3, and Figure 2-4 were having 32 segments.

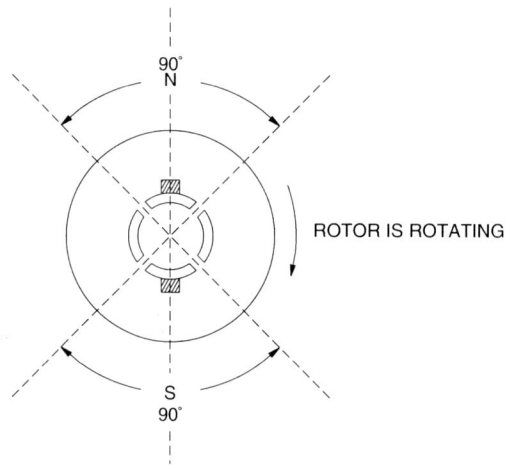

Figure 2-4. The north and south poles oscillate around a fixed position.

If this rotor is placed next to a fixed permanent magnet stator as shown in Figure 2-5, the magnetic poles of opposite polarity attract each other (in order to align) and the rotor starts to turn. After the rotor has turned by a certain angle, a commutation occurs and the north and south poles of the electromagnet go back. Once again, the magnetic poles of opposite polarity attract each other, and the rotor continues to rotate in the same direction so as to align the magnetic poles of opposite polarity. However, another commutation occurs a little after and the north and south poles of the electromagnet go back once again. This cycle repeats over and over. The force that results from the interaction of the two magnetic fields always acts in the same direction, and the rotor turns continually. Thus, a converter of electrical-to-mechanical energy, i.e., an electric motor, has been achieved. The direction of rotation depends on the polarity of the voltage applied to the brushes of the rotor.

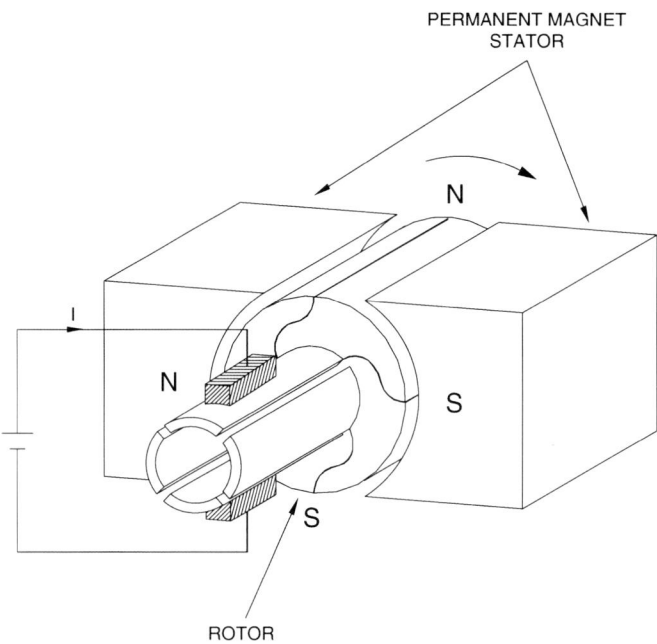

Figure 2-5. Rotation resulting from interaction of magnetic fields in the stator and the rotor.

Operating principle of dc generators

Previously, we saw that the variation of magnetic flux in a coil of wire caused a voltage to be induced between the ends of the coil of wire. If a wire loop is placed between two magnets and rotated as shown in Figure 2-6, magnetic lines of force are cut and a voltage "e" is induced in the loop. The polarity of the induced voltage "e" depends on the direction in which the wire loop moves as it cuts the magnetic lines of force. Since the wire loop cuts magnetic lines of force in both directions within a full revolution, the induced voltage is an ac voltage similar to that shown in Figure 2-6.

If a commutator such as that shown in Figure 2-1 is used, it will act as a **rectifier** and convert the induced ac voltage into a dc voltage (with ripple), as shown in Figure 2-6. Direct current will therefore be produced at the output of the generator. The faster the rotor turns, the more lines of force that are cut and the higher the output voltage. Also, the stronger the stator magnet, the more lines of force that are present, and therefore, the higher the output voltage.

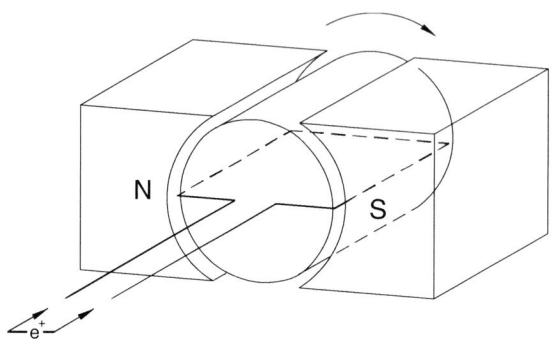

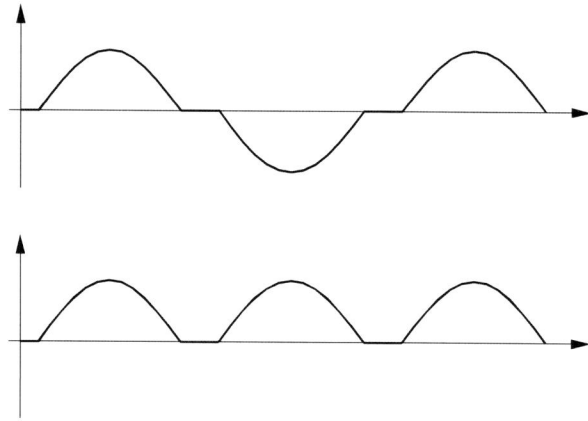

Figure 2-6. A coil rotating in a magnetic field results in an induced voltage.

Exercise 2-1

The Separately-Excited DC Motor

EXERCISE OBJECTIVE When you have completed this exercise, you will be able to demonstrate the main operating characteristics of a separately-excited dc motor using the DC Motor/Generator module.

DISCUSSION Previously, you saw that a dc motor is made up basically of a fixed magnet (stator) and a rotating magnet (rotor). Many dc motors use an electromagnet for the stator, as illustrated in Figure 2-7.

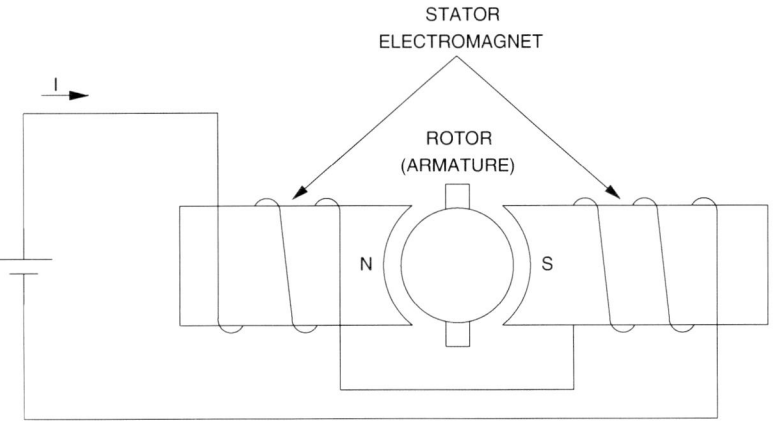

Figure 2-7. Simplified dc motor with an electromagnet as stator.

When power for the stator electromagnet is supplied by a separate dc source, either fixed or variable, the motor is known as a separately-excited dc motor. Sometimes the term independent-field dc motor is also used. The current flowing in the stator electromagnet is often called **field current** because it is used to create a fixed magnetic field. The electrical and mechanical behaviour of the dc motor can be understood by examining its simplified equivalent electric circuit shown in Figure 2-8.

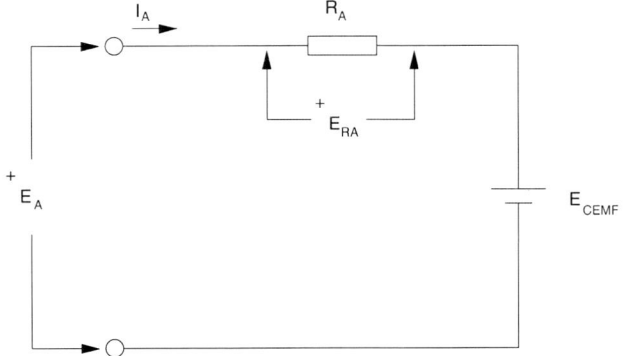

Figure 2-8. Simplified equivalent circuit of a dc motor.

In the circuit, E_A is the voltage applied to the motor brushes, I_A is the current flowing through the brushes, and R_A is the resistance between the brushes. Note that E_A, I_A, and R_A are usually referred to as the armature voltage, current, and resistance, respectively. E_{RA} is the voltage drop across the armature resistor. When the motor turns, an induced voltage E_{CEMF} proportional to the speed of the motor is produced. This induced voltage is represented by a dc source in the simplified equivalent circuit of Figure 2-8. The motor also develops a torque T proportional to the armature current I_A flowing in the motor. The motor behaviour is based on the two equations given below. The first relates motor speed n and the induced voltage E_{CEMF}, and the second relates the motor torque T and the armature current I_A.

$$n = K_1 \times E_{CEMF} \quad \text{and} \quad T = K_2 \times I_A$$

where K_1 is a constant expressed in units of r/min / V.

K_2 is a constant expressed in units of N·m / A or lbf·in / A.

When a voltage E_A is applied to the armature of a dc motor with no mechanical load, the armature current I_A flowing in the equivalent circuit of Figure 2-8 is constant and has a very low value. As a result, the voltage drop E_{RA} across the armature resistor is so low that it can be neglected, and E_{CEMF} can be considered to be equal to the armature voltage E_A. Therefore, the relationship between the motor speed n and the armature voltage E_A is a straight line because E_{CEMF} is proportional to the motor speed n. This linear relationship is illustrated in Figure 2-9, and the slope of the straight line equals constant K_1.

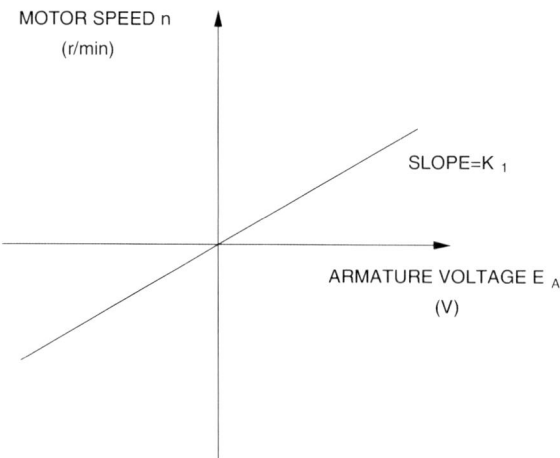

Figure 2-9. Linear relationship between the motor speed and the armature voltage.

Since the relationship between voltage E_A and speed n is linear, a dc motor can be considered to be a linear voltage-to-speed converter, as shown in Figure 2-10.

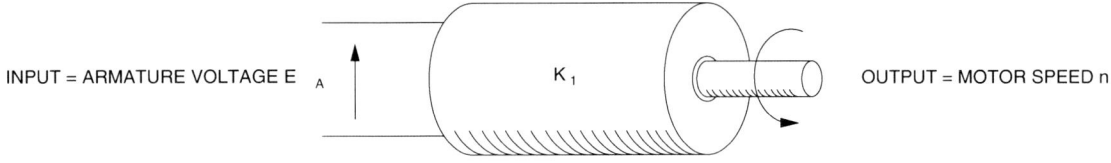

Figure 2-10. DC motor as a voltage-to-speed converter.

The same type of relationship exists between the motor torque T and the armature current I_A, so that a dc motor can also be considered as a linear current-to-torque converter. Figure 2-11 illustrates the linear relationship between the motor torque T and the armature current I_A. Constant K_2 is the slope of the line relating the two. Figure 2-12 shows the linear current-to-torque converter.

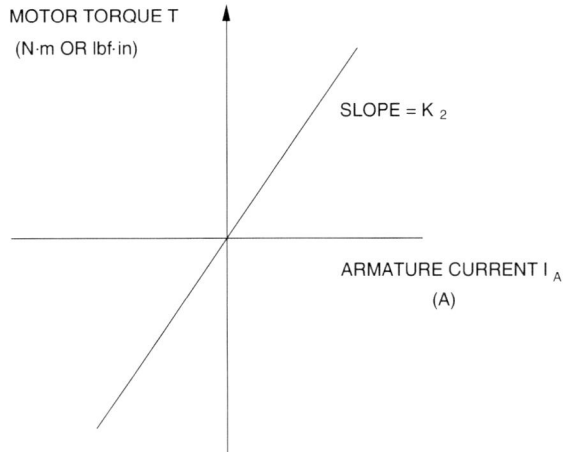

Figure 2-11. Linear relationship between the motor torque and the armature current.

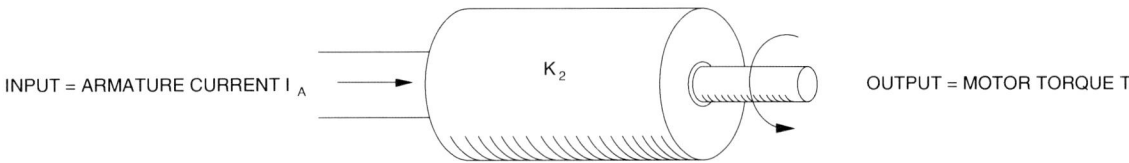

Figure 2-12. DC motor as a current-to-torque converter.

When the armature current I_A increases, the voltage drop E_{RA} ($R_A \times I_A$) across the armature resistor also increases and can no longer be neglected. As a result, the armature voltage E_A can no longer be considered to be equal to E_{CEMF}, but rather the sum of E_{CEMF} and E_{RA}, as indicated in the following equation:

$$E_A = E_{CEMF} + E_{RA}$$

Therefore, when a fixed armature voltage E_A is applied to a dc motor, the voltage drop E_{RA} across the armature resistor increases as the armature current I_A increases, and thereby, causes E_{CEMF} to decrease. This also causes the motor speed n to decrease because it is proportional to E_{CEMF}. This is shown in Figure 2-13, which is a graph of the motor speed n versus the armature current I_A for a fixed armature voltage E_A.

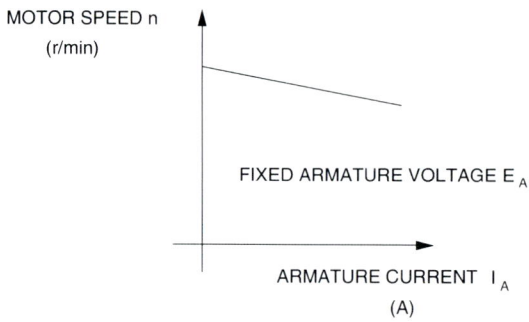

Figure 2-13. Motor speed drop as the armature current increases (fixed armature voltage E_A).

Procedure summary

In the first part of the exercise, you will set up the equipment in the Workstation, connect the equipment as shown in Figure 2-14, and make the appropriate settings on the equipment.

In the second part of the exercise, you will measure the armature resistance R_A of the DC Motor/Generator. It is not possible to measure the armature resistance R_A directly with a conventional ohmmeter because the non-linear characteristic of the motor brushes causes incorrect results when I_A is too small. The general method used to determine the armature resistance R_A consists in connecting a dc power source to the motor armature and measuring the voltage required to produce nominal current flow in the armature windings. Power is not connected to the stator electromagnet to ensure that the motor does not turn. Therefore, E_{CEMF} equals zero. The ratio of the armature voltage E_A to the armature current I_A yields the armature resistance R_A directly.

 The motor will not start to rotate because it is mechanically loaded.

In the third part of the exercise, you will measure data and plot a graph of the motor speed n versus the armature voltage E_A to demonstrate that the speed of the separately-excited dc motor is proportional to the armature voltage E_A under no-load conditions.

In the fourth part of the exercise, you will measure data and plot a graph of the motor torque T versus the armature current I_A to demonstrate that the torque of the separately-excited dc motor is proportional to the armature current I_A.

Ex. 2-1 – The Separately-Excited DC Motor ◆ *Procedure*

In the fifth part of the exercise, you will demonstrate that when the armature voltage E_A is set to a fixed value, the speed of the separately-excited dc motor decreases with increasing armature current or torque because of the increasing voltage drop across the armature resistor.

EQUIPMENT REQUIRED

Refer to the Equipment Utilization Chart in Appendix C to obtain the list of equipment required for this exercise.

PROCEDURE

 High voltages are present in this laboratory exercise. Do not make or modify any banana jack connections with the power on unless otherwise specified.

Setting up the equipment

1. Install the equipment required in the EMS workstation

    If you are performing the exercise using the EMS system, ensure that the brushes of the DC Motor/Generator are adjusted to the neutral point. To do so, connect an ac power source (terminals 4 and N of the Power Supply) to the armature of the DC Motor/Generator (terminals 1 and 2) through CURRENT INPUT I1 of the data acquisition module. Connect the shunt winding of the DC Motor/Generator (terminals 5 and 6) to VOLTAGE INPUT E1 of the data acquisition module. Start the Metering application and open setup configuration file ACMOTOR1.DAI. Turn the Power Supply on and set the voltage control knob so that an ac current (indicated by meter I line 1) equal to half the nominal value of the armature current flows in the armature of the DC Motor/Generator. Adjust the brush adjustment lever on the DC Motor/Generator so that the voltage across the shunt winding (indicated by meter E line 1) is minimum. Turn the Power Supply off, exit the Metering application, and disconnect all leads and cable.

   Mechanically couple the prime mover/dynamometer module to the DC Motor/Generator using a timing belt.

2. On the Power Supply, make sure the main power switch is set to the O (off) position, and the voltage control knob is turned fully counterclockwise. Ensure the Power Supply is connected to a three-phase power source.

    If you are using the Four-Quadrant Dynamometer/Power Supply, Model 8960-2, connect its POWER INPUT to a wall receptacle.

3. Ensure that the data acquisition module is connected to a USB port of the computer.

   Connect the POWER INPUT of the data acquisition module to the 24 V - AC output of the Power Supply.

    If you are using the Prime Mover/Dynamometer, Model 8960-1, connect its LOW POWER INPUT to the 24 V - AC output of the Power Supply.

On the Power Supply, set the 24 V - AC power switch to the I (on) position.

If you are using the Four-Quadrant Dynamometer/Power Supply, Model 8960-2, turn it on by setting its POWER INPUT switch to the I (on) position. Press and hold the FUNCTION button 3 seconds to have uncorrected torque values on the display of the Four-Quadrant Dynamometer/Power Supply. The indication "NC" appears next to the function name on the display to indicate that the torque values are uncorrected.

4. Start the Data Acquisition software (LVDAC or LVDAM). Open setup configuration file DCMOTOR1.DAI.

If you are using LVSIM-EMS in LVVL, you must use the IMPORT option in the File menu to open the configuration file.

In the Metering window, select layout 2. Make sure that the continuous refresh mode is selected.

5. Set up the separately-excited dc motor circuit shown in Figure 2-14. Leave the circuit open at points A and B shown in the figure.

6. Set the Four-Quadrant Dynamometer/Power Supply or the Prime Mover/Dynamometer to operate as a brake, then set the torque control to maximum (knob turned fully clockwise). To do this, refer to Exercise 1-1 or Exercise 1-2 if necessary.

If you are performing the exercise using LVSIM®-EMS, you can zoom in on the Prime Mover/Dynamometer module before setting the controls in order to see additional front panel markings related to these controls.

Ex. 2-1 – The Separately-Excited DC Motor ♦ *Procedure*

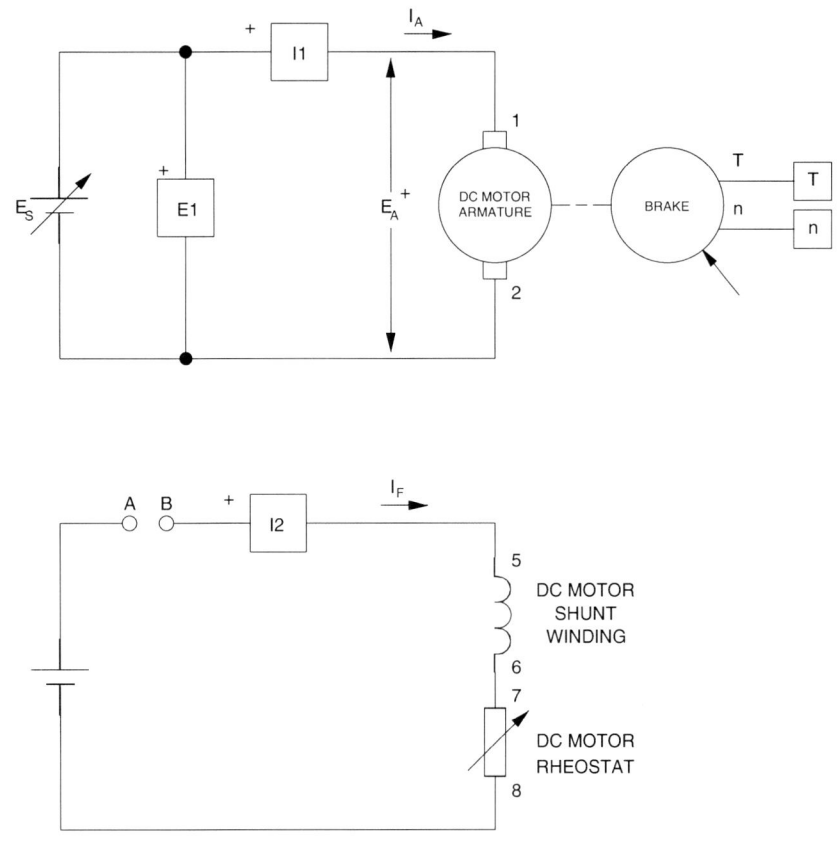

Figure 2-14. Separately-excited dc motor coupled to a brake.

Determining the armature resistance

7. Turn the Power Supply on by setting its main power switch to the I (on) position, and set the voltage control knob so that the rated armature current flows in the DC Motor/Generator. The armature current is indicated by meter I arm. (I_A) in the Metering window.

     The rating of any of the machines is indicated in the lower left corner of the module front panel.

    Record the value of armature resistance R_A indicated by meter $R_A = E_A/I_A$.

    $R_A = $ _____ Ω

8. Turn the voltage control knob fully counterclockwise and turn the Power Supply off.

    Interconnect points A and B shown in the circuit of Figure 2-14.

Ex. 2-1 – The Separately-Excited DC Motor ♦ Procedure

Motor speed versus armature voltage

9. Turn the Power Supply on.

   On the brake, set the torque control to minimum (knob turned fully counterclockwise).

   On the DC Motor/Generator, set the FIELD RHEOSTAT so that the field current I_f indicated by meter I field in the Metering window is equal to the value given in Table 2-1.

   Table 2-1. DC motor field current.

Local ac power network		I_f (mA)
Voltage (V)	Frequency (Hz)	
120	60	300
220	50	190
220	60	190
240	50	210

10. In the Metering window, make sure that the torque correction function of the Torque meter is enabled. The Torque meter now indicates the dc motor output torque. Record the armature voltage, armature current, field current, output torque, and speed in the Data Table. These parameters are indicated by meters E arm. (E_A), I arm. (I_A), I field (I_f), Torque, and Speed, respectively.

    On the Power Supply, set the voltage control knob to 10%, 20%, 30% etc. up to 100% in order to increase the armature voltage E_A by steps. For each voltage setting, wait until the motor speed stabilizes, and then record the data in the Data Table.

11. When all data has been recorded, turn the voltage control knob fully counterclockwise and turn the Power Supply off.

    In the Data Table window, confirm that the data has been stored, entitle the Data Table as DT211, and print the Data Table.

     Refer to the user guide dealing with the computer-based instruments for EMS to know how to edit, entitle, and print a Data Table.

Ex. 2-1 – The Separately-Excited DC Motor ◆ *Procedure*

12. In the Graph window, make the appropriate settings to obtain a graph of the dc motor speed n (obtained from the Speed meter) as a function of the armature voltage E_A [(obtained from meter E arm. (E_A)]. Entitle the graph as G211, name the x-axis as Armature voltage, name the y-axis as DC motor speed, and print the graph.

 Refer to the user guide dealing with the computer-based instruments for EMS to know how to use the Graph window of the Metering application to obtain a graph, entitle a graph, name the axes of a graph, and print a graph.

What kind of relationship exists between the armature voltage E_A and the dc motor speed n?

_____

Does this graph confirm that the separately-excited dc motor is equivalent to a linear voltage-to-speed converter, with higher voltage producing greater speed?

❏ Yes   ❏ No

13. Use the two end points to calculate the slope K_1 of the relationship obtained in graph G211. The values of these points are indicated in Data Table DT211.

$$K_1 = \frac{n_2 - n_1}{E_2 - E_1} = \frac{-}{-} = \frac{\text{r/min}}{\text{V}}$$

In the Data Table window, clear the recorded data.

Motor torque versus armature current

14. Turn the Power Supply on.

On the DC Motor/Generator, slightly readjust the FIELD RHEOSTAT so that the field current I_f indicated by meter I field (I_f) still equals the value given in Table 2-1 (if necessary).

On the Power Supply, set the voltage control knob so that the dc motor speed is 1500 r/min. Note the value of the armature voltage E_A in the following blank space.

$E_A = $ _____ V $(n = 1500 \text{ r/min})$

15. Record the dc motor armature voltage E_A, armature current I_A, field current I_f, output torque T, and speed n in the Data Table.

Ex. 2-1 – The Separately-Excited DC Motor ♦ *Procedure*

On the brake, set the torque control so that the torque indicated by the Torque meter in the Metering window increases by 0.2 N·m (2.0 lbf·in) increments up to about 2.3 N·m (about 20.4 lbf·in). For each torque setting, readjust the voltage control knob of the Power Supply so that the armature voltage E_A remains equal to the value recorded in the previous step, then record the data in the Data Table.

The armature current may exceed the rated value while performing this manipulation. It is, therefore, suggested to complete the manipulation within a time interval of 5 minutes or less.

16. When all data has been recorded, set the torque control on the brake to minimum (knob turned fully counterclockwise), turn the voltage control knob fully counterclockwise, and turn the Power Supply off.

In the Data Table window, confirm that the data has been stored, entitle the Data Table as DT212, and print the Data Table.

17. In the Graph window, make the appropriate settings to obtain a graph of the dc motor torque (obtained from the Torque meter) as a function of the armature current I_A [obtained from meter I arm. (I_A)]. Entitle the graph as G212, name the x-axis as Armature current, name the y-axis as DC motor torque, and print the graph.

What kind of relationship exists between the armature current I_A and the dc motor torque T as long as the armature current does not exceed the nominal value?

_____

Does this graph confirm that the separately-excited dc motor is equivalent to a linear current-to-torque converter (when the armature current does not exceed the nominal value), with higher current producing greater torque?

❑ Yes   ❑ No

The torque versus current relationship is no longer linear when the armature current exceeds the nominal value because of a phenomenon called armature reaction. This phenomenon is described in the next unit of this manual.

18. Use the two end points of the linear portion of the relationship obtained in graph G212 to calculate the slope K_2. The values of these points are indicated in Data Table DT212.

$$K_2 = \frac{T_2 - T_1}{I_2 - I_1} = \frac{-}{-} = \frac{}{} \frac{\text{N} \cdot \text{m (lbf} \cdot \text{in)}}{\text{A}}$$

Ex. 2-1 – The Separately-Excited DC Motor ◆ *Procedure*

Speed decrease versus armature current

19. Using the armature resistance R_A and the constant K_1 determined previously in this exercise, the armature voltage E_A measured in step 14, and the set of equations given below, determine the dc motor speed n for each of the three armature currents I_A given in Table 2-2.

$$E_{RA} = I_A \times R_A$$

$$E_{CEMF} = E_A - E_{RA}$$

$$n = E_{CEMF} \times K_1$$

Table 2-2. DC motor armature currents.

Local ac power network		Armature current I_A (A)	Armature current I_A (A)	Armature current I_A (A)
Voltage (V)	Frequency (Hz)			
120	60	1.0	2.0	3.0
220	50	0.5	1.0	1.5
220	60	0.5	1.0	1.5
240	50	0.5	1.0	1.5

When $I_A = $ _____ A:

$E_{RA} = $ _____ V

$E_{CEMF} = $ _____ V

$n = $ _____ r/min

When $I_A = $ _____ A:

$E_{RA} = $ _____ V

$E_{CEMF} = $ _____ V

$n = $ _____ r/min

When $I_A =$ _____ A:

$E_{RA} =$ _____ V

$E_{CEMF} =$ _____ V

$n =$ _____ r/min

Based on your results, how should the dc motor induced voltage E_{CEMF} and speed n vary when the armature current I_A is increased?

_____

_____

_____

20. In the Graph window, make the appropriate settings to obtain a graph of the dc motor speed (obtained from the Speed meter) as a function of the armature current I_A [obtained from meter I arm. (I_A)], using the data recorded previously in the Data Table (DT212). Entitle the graph as G212-1, name the x-axis as Armature current, name the y-axis as DC motor speed, and print the graph.

Does graph G212-1 confirm the prediction you made in the previous step about the variation of the dc motor speed as a function of the armature current I_A?

❑ Yes   ❑ No

Briefly explain what causes the dc motor speed to decrease when the armature voltage E_A is fixed and the armature current I_A increases.

_____

_____

_____

21. In the Graph window, make the appropriate settings to obtain a graph of the dc motor speed (obtained from the Speed meter) as a function of the dc motor torque T (obtained from the Torque meter), using the data recorded previously in the Data Table (DT212). Entitle the graph as G212-2, name the x-axis as Separately-excited dc motor Torque, name the y-axis as Separately-excited dc motor speed, and print the graph. This graph will be used in the next exercise of this unit.

22. On the Power Supply, set the 24 V - AC power switch to the O (off) position.

 If you are using the Four-Quadrant Dynamometer/Power Supply, Model 8960-2, turn it off by setting its POWER INPUT switch to the O (off position).

Remove all leads and cables.

Ex. 2-1 – The Separately-Excited DC Motor ♦ Conclusion

Additional experiments

DC motor speed-versus-armature voltage graph and torque-versus-armature current graph for reversed armature connections

You can obtain graphs of the dc motor speed n versus the armature voltage E_A, and dc motor torque T versus the armature current I_A, with reversed armature connections. To do so, make sure the Power Supply is turned off and reverse the connection of the leads at terminals 7 and N of the Power Supply. Refer to steps 6 to 17 of this exercise to record the necessary data and obtain the graphs. This will allow you to verify that the linear relationships between the speed and armature voltage, and the torque and armature current, are valid regardless of the polarity of the armature voltage. Recalculating constants K_1 and K_2 will show you that their values are independent of the armature voltage polarity.

CONCLUSION

In this exercise, you have learned how to measure the armature resistance of a dc motor. You have seen that the speed of a separately-excited dc motor is proportional to the armature voltage applied to the motor. You saw that the torque produced by a dc motor is proportional to the armature current. You observed that the dc motor speed decreases with increasing armature current when the armature voltage is fixed. You demonstrated that this speed decrease is caused by the increasing voltage drop across the armature resistor as the armature current increases.

If you have performed the additional experiments, you observed that the speed versus voltage and torque versus current relationships are not affected by the polarity of the armature voltage. You also observed that the direction of rotation is reversed when the polarity of the armature voltage is reversed.

REVIEW QUESTIONS

1. What kind of relationship exists between the speed and armature voltage of a separately-excited dc motor?

    a. A linear relationship
    b. A parabolic relationship.
    c. An exponential relationship.
    d. The speed of the motor is independent of the applied voltage.

2. What kind of relationship exists between the torque and armature current of a separately-excited dc motor as long as the armature current does not exceed the nominal value?

    a. A linear relationship.
    b. A parabolic relationship.
    c. An exponential relationship.
    d. The motor torque is independent of the current.

3. Connecting a dc source to the armature of a dc motor that operates without field current and measuring the voltage that produces nominal current flow in the armature allows which parameter of the dc motor to be determined?

   a. The nominal armature current.
   b. The nominal armature voltage.
   c. The armature resistance.
   d. The resistance of the field winding.

4. Does the speed of a separately-excited dc motor increase or decrease when the armature current increases?

   a. It increases.
   b. It decreases.
   c. It stays the same because speed is independent of motor current.
   d. The speed will oscillate around the previous value.

5. The armature resistance R_A and constant K_1 of a dc motor are 0.5 Ω and 5 r/min/V, respectively. A voltage of 200 V is applied to this motor. The no-load armature current is 2 A. At full load, the armature current increases to 50 A. What are the no-load and full-load speeds of the motor?

   a. $n_{NO\ LOAD}$ = 1005 r/min, $n_{FULL\ LOAD}$ = 880 r/min
   b. $n_{NO\ LOAD}$ = 995 r/min, $n_{FULL\ LOAD}$ = 875 r/min
   c. $n_{NO\ LOAD}$ = 1000 r/min, $n_{FULL\ LOAD}$ = 875 r/min
   d. The speeds cannot be calculated without constant K_2.

Exercise 2-2

Separately-Excited, Series, Shunt, and Compound DC Motors

EXERCISE OBJECTIVE When you have completed this exercise, you will be able to demonstrate how the field current affects the characteristics of a separately-excited dc motor using the DC Motor/Generator module. You will also be able to demonstrate the main operating characteristics of series, shunt, and compound motors.

DISCUSSION

Separately-excited dc motor

It is possible to change the characteristics of a separately-excited dc motor by changing the strength of the fixed magnetic field produced by the stator electromagnet. This can be carried out by changing the current that flows in the stator electromagnet. This current is usually referred to as the field current (I_f) because it is used to produce the fixed magnetic field in the dc motor. A rheostat connected in series with the electromagnet winding can be used to vary the field current.

Figure 2-15 illustrates how the speed versus armature voltage and torque versus armature current relationships of a separately-excited dc motor are affected when the field current is decreased below its nominal value. Constant K_1 becomes greater and constant K_2 becomes smaller. This means that the motor can rotate at higher speeds without exceeding the nominal armature voltage. However, the torque which the motor can develop, without exceeding the nominal armature current, is reduced.

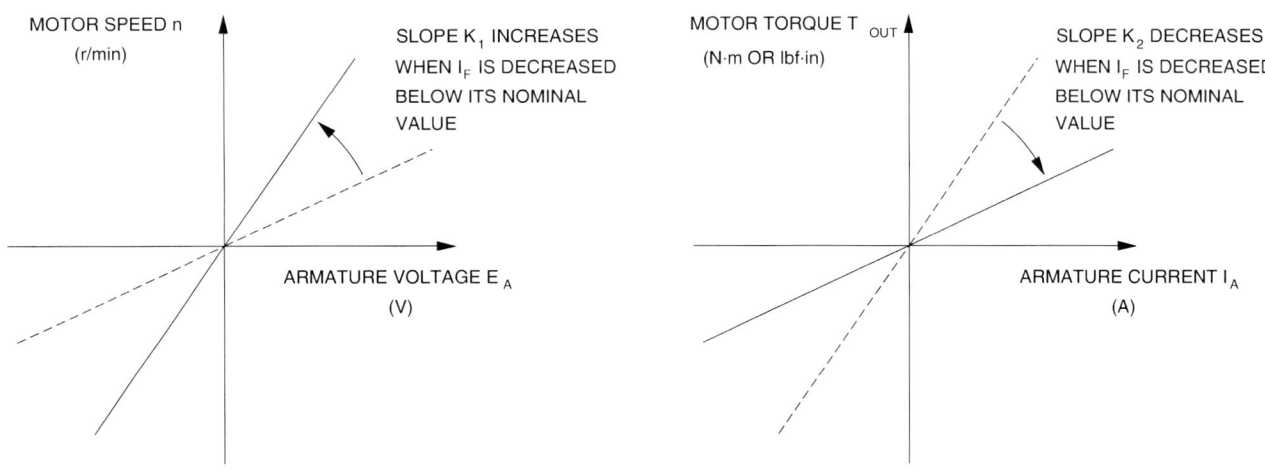

Figure 2-15. Decreasing current I_f below its nominal value affects constants K_1 and K_2.

It is also possible to set the field current of a separately-excited dc motor above its nominal value for short time intervals. The effect on the speed versus armature voltage and torque versus armature current relationships is reversed, i.e., constant K_1 becomes smaller and constant K_2 becomes higher. As a result, the motor can develop a higher torque during these time intervals but the speed at which the motor can rotate, without exceeding the nominal armature voltage, is reduced. Increasing the field current of a separately-excited dc motor when it is starting improves the motor torque, and thereby, provides faster acceleration.

The strength of the fixed magnetic field in a dc motor can also be changed by changing the way the stator electromagnet is implemented. The stator electromagnet, or field electromagnet, can be a shunt winding connected directly to a dc voltage source, as in the separately-excited dc motor. A shunt winding can also be connected in parallel with the armature of the dc motor. The field electromagnet can also be a series winding, a coil consisting of a few loops of heavy-gage wire, connected in series with the armature. A combination of the shunt and series windings can also be used to implement the field electromagnet.

Various electromagnet implementations have been used so far to build several types of dc motors having different characteristics when powered by a fixed-voltage dc source. This was necessary at the time the first dc motors were in used, because variable-voltage dc sources were not still available. These dc motors, which are used less and less today, are briefly described in the following sections of this discussion.

Series motor

The series motor is a motor in which the field electromagnet is a series winding connected in series with the armature as shown in Figure 2-16. The strength of the field electromagnet, therefore, varies as the armature current varies. As a result, K_1 and K_2 vary when the armature current varies. Figure 2-16 shows the speed versus torque characteristic of a series motor when the armature voltage is fixed. This characteristic shows that the speed decreases non linearly as the torque increases, i.e., as the armature current increases.

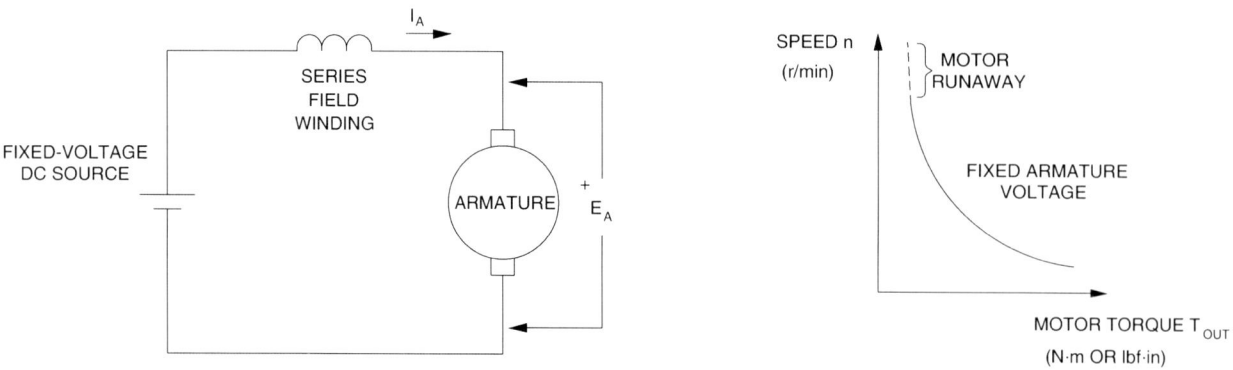

Figure 2-16. Series motor and its speed versus torque characteristic.

The series motor provides a strong starting torque and a wide range of operating speeds when it is supplied by a fixed-voltage dc source. However, the speed, torque, and armature current depend on the mechanical load applied to the motor. Also, the series motor has non-linear operating characteristics as suggested by the speed versus torque relationship in Figure 2-16. As a result, it is difficult to operate a series motor at a constant speed when the mechanical load fluctuates. Furthermore, the armature current must be limited to prevent damage to the motor when it is starting (when power is applied to the motor). Finally, a series motor must never run without mechanical load because the speed increases to a very-high value which can damage the motor (motor runaway).

Today, series motors can operate with fixed-voltage power sources, for example, automobile starting motors; or with variable-voltage power sources, for example, traction systems.

Shunt motor

The shunt motor is a motor in which the field electromagnet is a shunt winding connected in parallel with the armature, both being connected to the same dc voltage source as shown in Figure 2-17. For a fixed armature voltage, constants K_1 and K_2 are fixed, and the speed versus torque characteristic is very similar to that obtained with a separately-excited dc motor powered by a fixed-voltage dc source, as shown in Figure 2-17. As in a separately-excited dc motor, the characteristics (K_1 and K_2) of a shunt motor can be changed by varying the field current with a rheostat. However, it is difficult to change the speed of a shunt motor by changing the armature voltage, because this changes the field current, and thereby, the motor characteristics, in a way that opposes speed change.

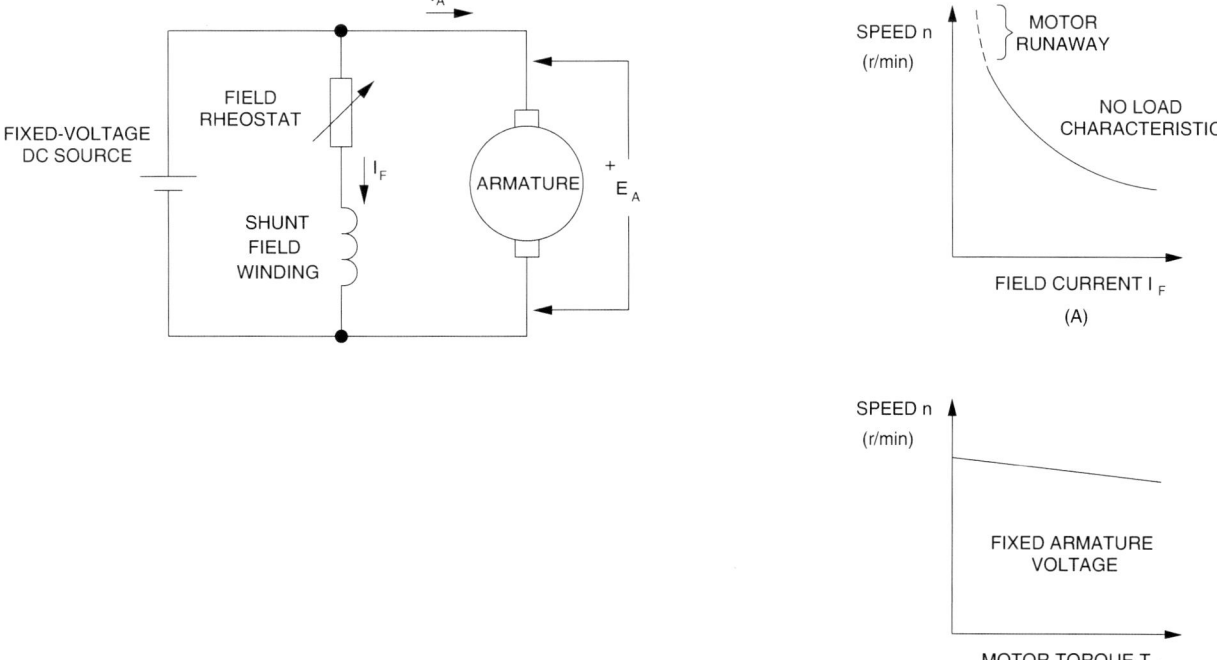

Figure 2-17. Shunt motor and its characteristics.

The main advantage of a shunt motor is the fact that only a single fixed-voltage dc source is required to supply power to both the armature and the shunt winding. Also, speed varies little as the mechanical load varies. However, a shunt motor has a limited speed range because speed cannot be easily varied by varying the armature voltage. Furthermore, the armature current must be limited to prevent damage to the motor when it is starting (when power is applied to the motor). Finally, when the shunt winding opens accidentally, the field current I_f becomes zero, the motor speed increases rapidly, and motor runaway occurs as suggested by the speed versus field current characteristic shown in Figure 2-17.

Compound motor

It is possible to combine shunt and series windings to obtain a particular speed versus torque characteristic. For example, to obtain the characteristic of decreasing speed when the motor torque increases, a series winding can be connected in series with the armature so that the magnetic flux it produces adds with the magnetic flux produced by a shunt winding. As a result, the magnetic flux increases automatically with increasing armature current. This type of dc motor is referred to as a cumulative compound motor because the magnetic fluxes produced by the series and shunt windings add together. Shunt and series windings can also be connected so that the magnetic fluxes subtract from each other. This connection produces a differential-compound motor, which is rarely used because the motor becomes unstable when the armature current increases. Figure 2-18 shows a compound motor and its speed versus torque characteristic (cumulative compound).

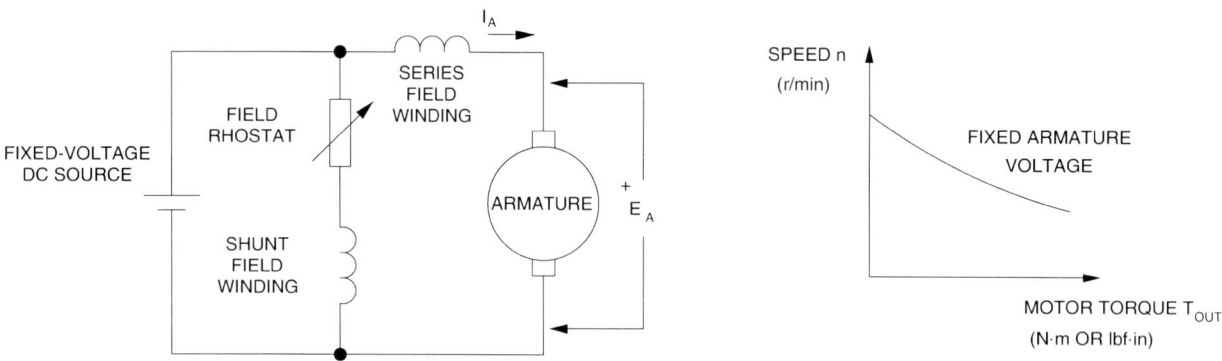

Figure 2-18. Compound motor and its speed versus torque characteristic.

Figure 2-19 is a graph that shows the speed versus torque characteristics of the various types of dc motors discussed so far. As can be seen, the separately-excited dc motor and the shunt motor have very similar characteristics. The main feature of these characteristics is that the motor speed varies little and linearly as the torque varies. On the other hand, the series motor characteristic is nonlinear and shows that the motor speed varies a lot (wide range of operating speed) as the torque varies. Finally, the characteristic of a cumulative compound motor is a compromise of the series and shunt motor characteristics. It provides the compound motor with a fairly wide range of operating speed, but the speed does not vary linearly as the torque varies.

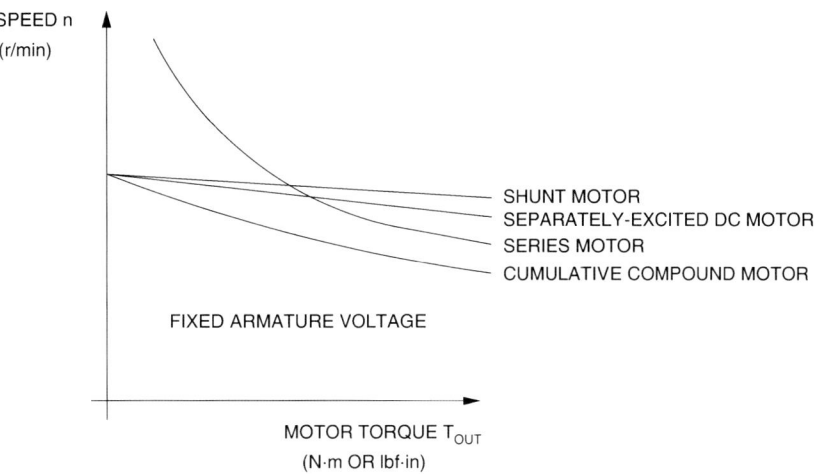

Figure 2-19. Speed versus torque characteristics of various dc motors.

Procedure summary

In the first part of the exercise, you will set up the equipment in the Workstation, connect the equipment as shown in Figure 2-20, and make the appropriate settings on the equipment.

In the second part of the exercise, you will set the field current of the separately-excited dc motor to a lower value than in the previous exercise (below the nominal value). You will measure data and plot a graph of the motor speed n versus the armature voltage E_A. You will calculate constant K_1. You will compare constant K_1 and the graph with those obtained in the previous exercise to determine how decreasing the field current affects these characteristics.

In the third part of the exercise, you will measure data and plot a graph of the motor torque T versus the armature current I_A. You will calculate constant K_2. You will compare constant K_2 and the graph with those obtained in the previous exercise to determine how decreasing the field current affects these characteristics.

In the fourth part of the exercise, you will connect the DC Motor/Generator as a series motor (see setup in Figure 2-21). You will measure data and plot a graph of the motor speed n versus the motor torque T. You will compare the speed versus torque characteristic of the series motor to that of the separately-excited dc motor obtained in the previous exercise.

EQUIPMENT REQUIRED

Refer to the Equipment Utilization Chart in Appendix C to obtain the list of equipment required for this exercise.

PROCEDURE

High voltages are present in this laboratory exercise. Do not make or modify any banana jack connections with the power on unless otherwise specified.

Setting up the equipment

1. Install the equipment required in the EMS workstation

    If you are performing the exercise using the EMS system, ensure that the brushes of the DC Motor/Generator are adjusted to the neutral point. To do so, connect an ac power source (terminals 4 and N of the Power Supply) to the armature of the DC Motor/Generator (terminals 1 and 2) through CURRENT INPUT I1 of the data acquisition module. Connect the shunt winding of the DC Motor/Generator (terminals 5 and 6) to VOLTAGE INPUT E1 of the data acquisition module. Start the Metering application and open setup configuration file ACMOTOR1.DAI. Turn the Power Supply on and set the voltage control knob so that an ac current (indicated by meter I line 1) equal to half the nominal value of the armature current flows in the armature of the DC Motor/Generator. Adjust the brush adjustment lever on the DC Motor/Generator so that the voltage across the shunt winding (indicated by meter E line 1) is minimum. Turn the Power Supply off, exit the Metering application, and disconnect all leads and cable.

    Mechanically couple the prime mover/dynamometer module to the DC Motor/Generator using a timing belt.

2. On the Power Supply, make sure the main power switch is set to the O (off) position, and the voltage control knob is turned fully counterclockwise. Ensure the Power Supply is connected to a three-phase power source.

    If you are using the Four-Quadrant Dynamometer/Power Supply, Model 8960-2, connect its POWER INPUT to a wall receptacle.

3. Ensure that the data acquisition module is connected to a USB port of the computer.

Connect the POWER INPUT of the data acquisition module to the 24 V - AC output of the Power Supply.

 If you are using the Prime Mover/Dynamometer, Model 8960-1, connect its LOW POWER INPUT to the 24 V - AC output of the Power Supply.

On the Power Supply, set the 24 V - AC power switch to the I (on) position.

 If you are using the Four-Quadrant Dynamometer/Power Supply, Model 8960-2, turn it on by setting its POWER INPUT switch to the I (on) position. Press and hold the FUNCTION button 3 seconds to have uncorrected torque values on the display of the Four-Quadrant Dynamometer/Power Supply. The indication "NC" appears next to the function name on the display to indicate that the torque values are uncorrected.

4. Start the Data Acquisition software (LVDAC or LVDAM). Open setup configuration file DCMOTOR1.DAI.

 If you are using LVSIM-EMS in LVVL, you must use the IMPORT option in the File menu to open the configuration file.

In the Metering window, select layout 1. Make sure that the continuous refresh mode is selected.

5. Set up the separately-excited dc motor circuit shown in Figure 2-20. Note that this setup is the same as that used in the previous exercise.

 If you are performing the exercise with a line voltage of 220 V, use the Resistive Load module to connect a 880-Ω resistor in series with the rheostat of the DC Motor/Generator. If you are performing the exercise with a line voltage of 240 V, connect a 960-Ω resistor in series with the rheostat.

Ex. 2-2 – Separately-Excited, Series, Shunt, and Compound DC Motors ♦ *Procedure*

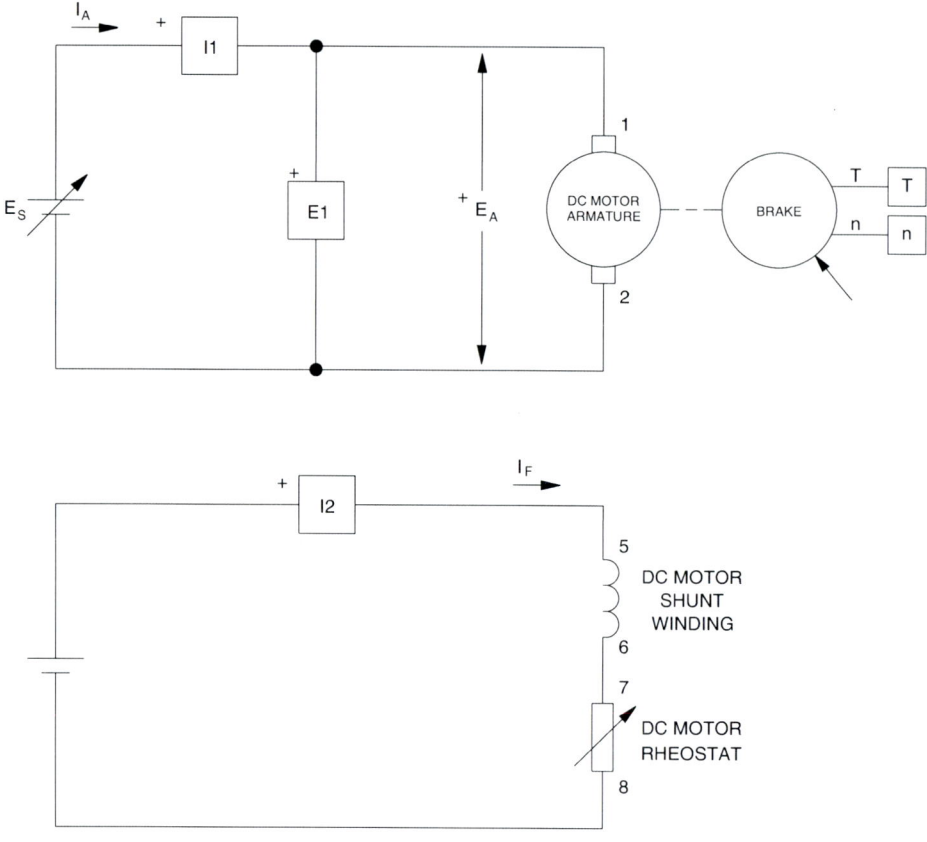

Figure 2-20. Separately-excited dc motor coupled to a brake.

6. Set the Four-Quadrant Dynamometer/Power Supply or the Prime Mover/Dynamometer to operate as a brake, then set the torque control to minimum (knob turned fully counterclockwise). To do this, refer to Exercise 1-1 or Exercise 1-2 if necessary.

 If you are performing the exercise using LVSIM®-EMS, you can zoom in on the Prime Mover/Dynamometer before setting the controls in order to see additional front panel markings related to these controls.

Speed versus armature voltage characteristic of a separately-excited dc motor

7. Turn the Power Supply on.

   On the DC Motor/Generator, set the FIELD RHEOSTAT so that the field current I_f indicated by meter I field (I_f) in the Metering window is equal to the value given in Table 2-3:

   Table 2-3. Field current of the separately-excited dc motor.

Local ac power network		I_f (mA)
Voltage (V)	Frequency (Hz)	
120	60	200
220	50	125
220	60	125
240	50	140

8. In the Metering window, make sure that the torque correction function of the Torque meter is enabled. The Torque meter now indicates the dc motor output torque. Record the armature voltage, armature current, field current, output torque, and speed in the Data Table. These parameters are indicated by meters E arm. (E_A), I arm. (I_A), I field (I_f), Torque, and Speed respectively.

   On the Power Supply, set the voltage control knob to 10%, 20%, 30%, etc. up to 100% in order to increase the armature voltage E_A by steps. For each voltage setting, wait until the motor speed stabilizes, and then record the data in the Data Table.

9. When all data has been recorded, turn the voltage control knob fully counterclockwise and turn the Power Supply off.

   In the Data Table window, confirm that the data has been stored, entitle the Data Table as DT221, and print the Data Table.

   Refer to the user guide dealing with the computer-based instruments for EMS to know how to edit, entitle, and print a Data Table.

10. In the Graph window, make the appropriate settings to obtain a graph of the dc motor speed n (obtained from the Speed meter) as a function of the armature voltage E_A [obtained from meter E arm. (E_A)]. Entitle the graph as G221, name the x-axis as Armature voltage, name the y-axis as DC motor speed, and print the graph.

    Refer to the user guide dealing with the computer-based instruments for EMS to know how to use the Graph window of the Metering application to obtain a graph, entitle a graph, name the axes of a graph, and print a graph.

11. Use the two end points to calculate the slope K_1 of the relationship obtained in graph G221. The values of these points are indicated in Data Table DT221.

$$K_1 = \frac{n_2 - n_1}{E_2 - E_1} = \frac{-}{-} = \underline{\hspace{1cm}} \frac{r/\text{min}}{V}$$

Compare graph G221 and constant K_1 obtained in this exercise with graph G211 and constant K_1 obtained in the previous exercise. How does decreasing the field current I_f affect the speed versus voltage characteristic and constant K_1 of a separately-excited dc motor?

_____

_____

_____

In the Data Table window, clear the recorded data.

Torque versus armature current characteristic of a separately-excited dc motor

12. Turn the Power Supply on.

On the DC Motor/Generator, slightly readjust the FIELD RHEOSTAT so that the field current I_f indicated by meter I field (I_f) still equals the value given in Table 2-3 (if necessary).

On the Power Supply, set the voltage control knob so that the dc motor speed is 1500 r/min. Note the value of the armature voltage E_A in the following blank space.

$E_A = $ _____ V   [$n = 1500$ r/min]

13. In the Metering window, make sure that the torque correction function of the Torque meter is enabled. Record the dc motor armature voltage E_A, armature current I_A, field current I_f, output torque T, and speed n in the Data Table.

On the brake, set the torque control so that the torque indicated by the Torque meter increases by 0.2 N·m (2 lbf·in) increments up to about 1.5 N·m (about 14.0 lbf·in). For each torque setting, readjust the voltage control knob of the Power Supply so that the armature voltage E_A remains equal to the value recorded in the previous step, then record the data in the Data Table.

The armature current may exceed the rated value while performing this manipulation. It is therefore suggested to complete the manipulation within a time interval of 5 minutes or less.

14. When all data has been recorded, set the torque on the brake to minimum (turned fully counterclockwise), turn the voltage control knob fully counterclockwise, and turn the Power Supply off.

In the Data Table window, confirm that the data has been stored, entitle the Data Table as DT222, and print the Data Table.

15. In the Graph window, make the appropriate settings to obtain a graph of the dc motor torque (obtained from the Torque meter) as a function of the armature current I_A [(obtained from meter I arm. (I_A)]. Entitle the graph as G222, name the x-axis as Armature current, name the y-axis as DC motor torque, and print the graph.

 The torque versus current relationship is no longer linear when the armature current exceeds the nominal value because of a phenomenon called armature reaction. This phenomenon is described in the next unit of this manual.

16. Use the two end points of the linear portion of the relationship obtained in graph G222 to calculate the slope K_2. The values of these points are indicated in Data Table DT222.

Use the two end points of the linear portion of the relationship obtained in graph G222 to calculate the slope K_2. The values of these points are indicated in Data Table DT222.

$$K_1 = \frac{T_2 - T_1}{I_2 - I_1} = \frac{-}{-} = \frac{\quad}{\quad} \frac{N \cdot m \, (lbf \cdot in)}{A}$$

Compare graph G222 and constant K_2 obtained in this exercise with graph G212 and constant K_2 obtained in the previous exercise. Describe how does decreasing the field current I_f affect the torque versus current characteristic and constant K_2 of a separately-excited dc motor.

___

___

___

In the Data Table window, clear the recorded data.

Speed versus torque characteristic of a series motor

17. Modify the connections so as to obtain the series motor circuit shown in Figure 2-21.

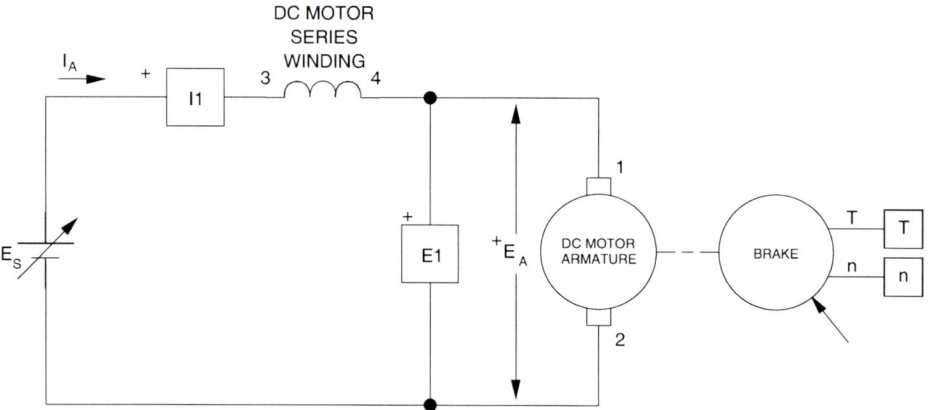

Figure 2-21. Series motor coupled to a brake.

18. Turn the Power Supply on and set the voltage control knob so that the armature voltage E_A indicated by meter E arm. (E_A) is equal to the value recorded in step 14 of the previous exercise. The series motor should start to rotate.

19. In the Metering window, make sure that the torque correction function of the Torque meter is enabled. Record the dc motor armature voltage E_A, armature current I_A, output torque T, and speed n in the Data Table.

On the dynamometer, set the torque control so that the torque indicated by the Torque meter increases by 0.2 N·m (2 lbf·in) increments up to about 2.3 N·m (about 20.3 lbf·in). For each torque setting, readjust the voltage control knob of the Power Supply so that the armature voltage E_A remains equal to the value set in the previous step, wait until the motor speed stabilizes, and then record the data in the Data Table.

 It may not be possible to maintain the armature voltage to its original value as the torque is increased. The armature current may exceed the rated value while performing this manipulation. It is, therefore, suggested to complete the manipulation within a time interval of 5 minutes or less.

20. When all data has been recorded, set the torque control on the brake to minimum (fully counterclockwise), turn the voltage control knob fully counterclockwise, and turn the Power Supply off.

In the Data Table window, confirm that the data has been stored, entitle the Data Table as DT223, and print the Data Table.

21. In the Graph window, make the appropriate settings to obtain a graph of the series motor speed (obtained from the Speed meter) as a function of the series motor torque (obtained from the Torque meter). Entitle the graph as G223, name the x-axis as Series motor torque, name the y-axis as Series motor speed, and print the graph.

Briefly describe how the speed varies as the mechanical load applied to the series motor increases, i.e., as the motor torque increases.

_____

_____

Compare the speed versus torque characteristic of the series motor (graph G223) to that of the separately-excited dc motor (graph G212-2 obtained in the previous exercise).

_____

_____

_____

22. On the Power Supply, set the 24 V - AC power switch to the O (off) position.

If you are using the Four-Quadrant Dynamometer/Power Supply, Model 8960-2, turn it off by setting its POWER INPUT switch to the O (off) position.

Remove all leads and cables.

Additional experiments

Speed versus torque characteristic of a shunt motor

You can obtain the speed versus torque characteristic of a shunt motor and compare it to those obtained for the separately-excited dc motor and series motor. To do so, make sure the Power Supply is turned off and set up the shunt motor circuit shown in Figure 2-22. (If you are using the Four-Quadrant Power Supply/Dynamometer, make sure to press and hold the FUNCTION button 3 seconds to have uncorrected values on this module's display). Make sure the torque control on the brake is set to minimum (fully counterclockwise). Turn the Power Supply on, set the armature voltage E_A to the value recorded in step 14 of the previous exercise. Set the FIELD RHEOSTAT on the DC Motor/Generator so that the field current I_f is equal to the value indicated in Table 2-1. Clear the data recorded in the Data Table. Refer to steps 19 to 21 of this exercise to record the necessary data and obtain the graph. Entitle the Data Table and graph as DT224 and G224, respectively. Compare the speed versus torque characteristic of the shunt motor (graph G224) to those of the separately-excited dc motor (graph G212-2 obtained in the previous exercise) and series motor (graph G223).

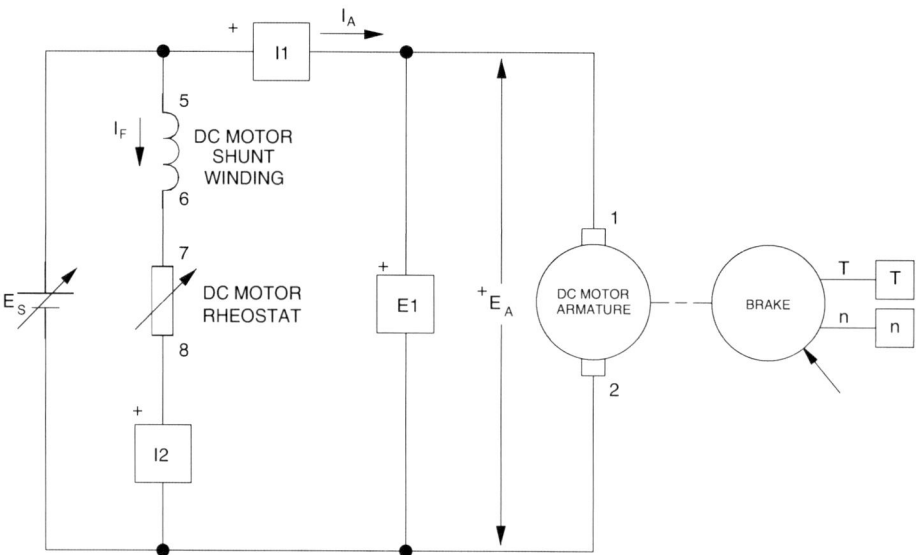

Figure 2-22. Shunt motor circuit.

Speed versus torque characteristic of a cumulative compound motor

You can obtain the speed versus torque characteristic of a cumulative compound motor and compare it to those obtained for the other dc motors. To do so, make sure the Power Supply is turned off and set up the cumulative compound motor circuit shown in Figure 2-23. (If you are using the Four-Quadrant Power Supply/Dynamometer, make sure to press and hold the FUNCTION button 3 seconds to have uncorrected values on this module's display). Make sure the torque control on the brake is set to minimum (fully counterclockwise). Turn the Power Supply on, set the armature voltage E_A to the value recorded in step 14 of the previous exercise. Set the FIELD RHEOSTAT on the DC Motor/Generator so that the current in the shunt winding I_f is equal to the value indicated in Table 2-1. Clear the data recorded in the Data Table. Refer to steps 19 to 21 of this exercise to record the necessary data and obtain the graph. Entitle the Data Table and graph as DT225 and G225, respectively. Compare the speed versus torque characteristic of the cumulative compound motor (graph G225) to those of the other dc motors (graphs G212-2, G223, and G224).

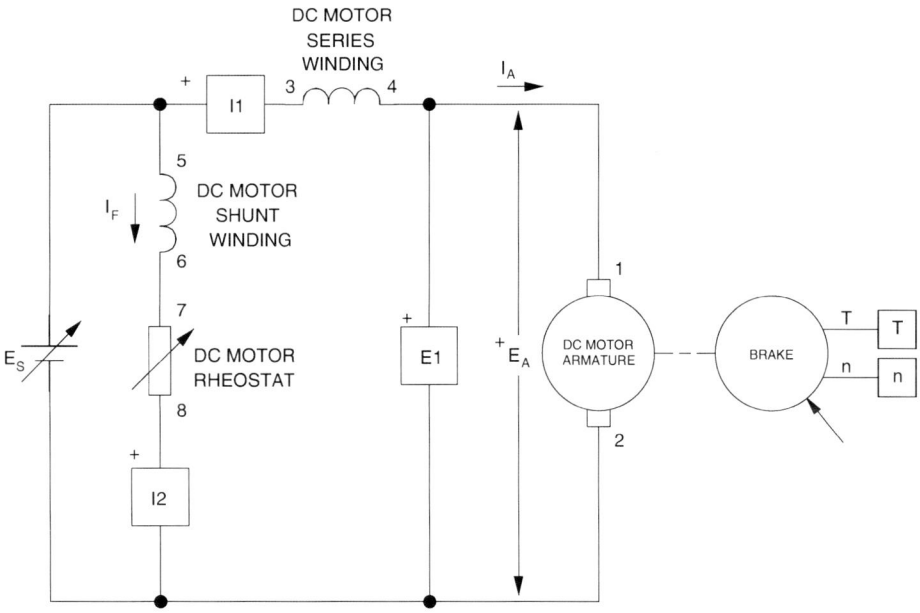

Figure 2-23. Cumulative compound motor circuit.

CONCLUSION

In this exercise, you observed that decreasing the field current of a separately-excited dc motor below its nominal value increases constant K_1 but decreases constant K_2. You saw that this allows the motor to rotate at higher speeds without exceeding the nominal armature voltage but reduces the torque which the motor can develop without exceeding the nominal armature current. You also saw that it is possible to increase the field current above its nominal value for short time intervals to improve the starting torque. You plotted a graph of the speed versus torque characteristic of a series motor and compared it to that obtained in the previous exercise with a separately-excited dc motor. You observed that the speed of a series motor decreases more rapidly than that of the separately-excited dc motor as the torque increases. Furthermore, you observed that the speed versus torque characteristic of the separately-excited dc motor is linear whereas that of the series motor is non linear.

If you have performed the additional experiments, you plotted graphs of the speed versus torque characteristic for a shunt motor and a cumulative compound motor. You compared these characteristics to those obtained with the separately-excited dc motor and the series motor. You found that the characteristic of a shunt motor is very similar to that of a separately-excited dc motor. You saw that the characteristic of a cumulative compound motor is a compromise of the characteristics of the separately-excited dc motor and series motor.

REVIEW QUESTIONS

1. What effect does decreasing the field current below its nominal value have on the speed versus voltage characteristic of a separately-excited dc motor?

   a. Constant K_1 increases.
   b. Constant K_2 increases.
   c. Constant K_1 decreases.
   d. Constant K_2 decreases.

2. What effect does decreasing the field current below its nominal value have on the torque-current characteristic of a separately-excited dc motor?

   a. Constant K_1 increases.
   b. Constant K_2 increases.
   c. Constant K_1 decreases.
   d. Constant K_2 decreases.

3. What is the advantage of increasing the field current above its nominal value for a short time interval when starting a separately-excited dc motor?

   a. This prevents damage to the motor.
   b. This allows the motor to reach a higher speed.
   c. This increases the armature voltage.
   d. This increases the starting torque.

4. Does the speed of a shunt motor increase or decrease when the armature current increases?

   a. It increases.
   b. It decreases.
   c. It oscillates around the previous value.
   d. It does not change because speed is independent of the armature current.

5. What is the advantage of decreasing the field current of a separately-excited dc motor below its nominal value?

   a. This allows the motor to develop a higher torque without exceeding the nominal armature voltage.
   b. This allows the motor to develop a higher torque without exceeding the nominal armature current.
   c. This allows the motor to rotate at a higher speed without exceeding the nominal armature voltage.
   d. This has no advantage.

Exercise 2-3

Separately-Excited, Shunt, and Compound DC Generators

EXERCISE OBJECTIVE

When you have completed this exercise, you will be able to demonstrate the main operating characteristics of separately-excited, shunt, and compound generators using the DC Motor/Generator module.

DISCUSSION

Although dc generators are rarely used today, it is important to know their operation because this helps understanding how a separately-excited dc motor can be used as an electric brake in modern dc motor drives.

You saw earlier in this unit that a dc motor can be considered as a linear voltage-to-speed converter. This linear conversion process is reversible, meaning that when a fixed speed is imposed on the motor by an external driving force, the motor produces an output voltage E_O, and thus, operates as a linear speed-to-voltage converter, i.e., a dc generator. Figure 2-24 illustrates a dc motor operating as a dc generator.

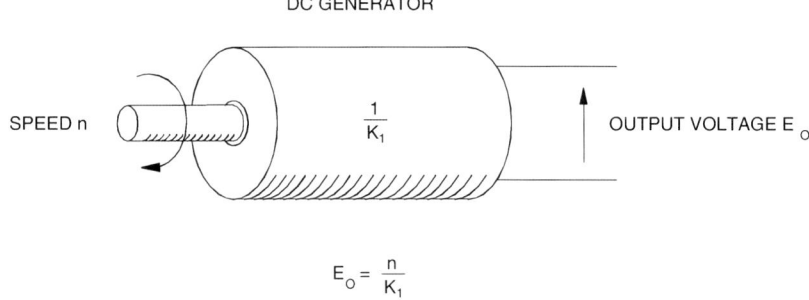

Figure 2-24. DC motor as a speed-to-voltage converter (dc generator).

The linear relationship that exists between torque and current for the dc motor is also reversible and applies to the dc generator, i.e., a torque must be applied to the generator's shaft to obtain a certain output current. Figure 2-25 illustrates a dc motor operating as a linear torque-to-current converter, i.e., a dc generator.

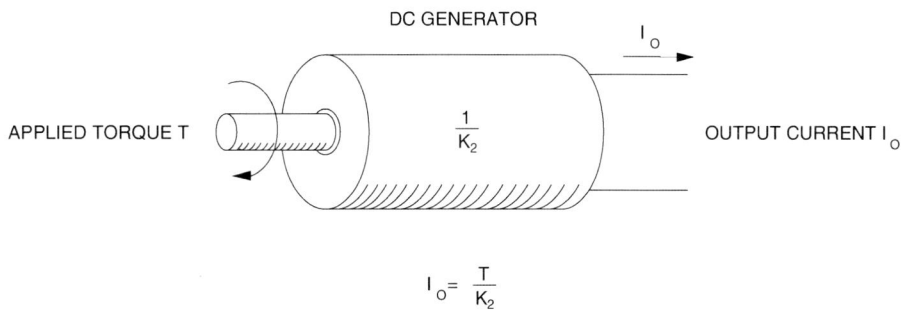

Figure 2-25. DC motor as a torque-to-current converter (dc generator).

Figure 2-26a shows the output voltage versus speed relationship of a separately-excited dc generator. Figure 2-26b shows the output current versus applied torque relationship of a separately-excited dc generator. Notice that the slopes of these linear relationships are equal to the reciprocal of constants K_1 and K_2.

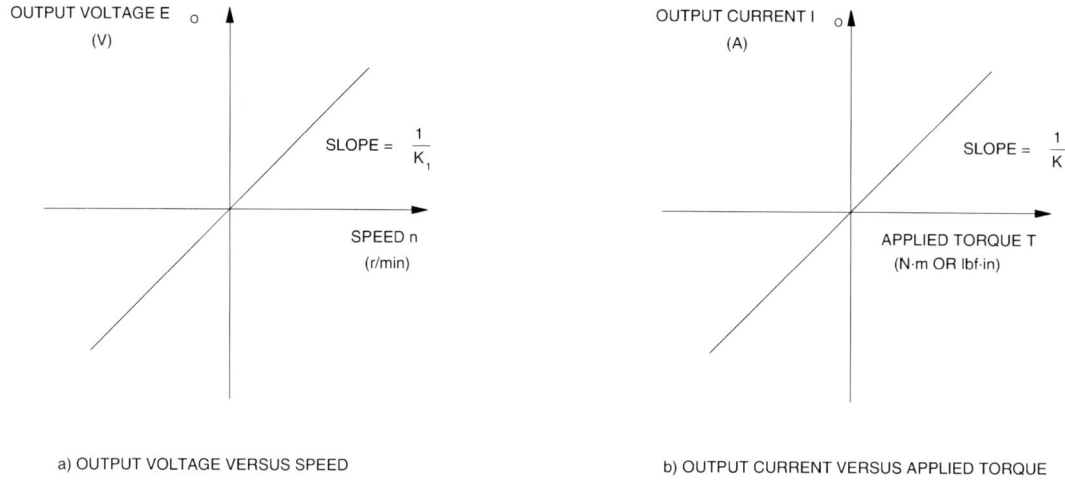

a) OUTPUT VOLTAGE VERSUS SPEED

b) OUTPUT CURRENT VERSUS APPLIED TORQUE

Figure 2-26. Input-output relationships of a separately-excited dc generator.

In a manner similar to that for a separately-excited dc motor, the field current I_f of a separately-excited dc generator can be varied to change the strength of the field electromagnet, and thereby, the relative values of constant K_1 and K_2. When the field current is decreased, constant K_1 increases and constant K_2 decreases, as for a separately-excited dc motor. As a result, the slope of the output voltage versus speed relationship decreases whereas the slope of the output current versus torque relationship increases. Conversely, when the field current is increased, constant K_1 decreases and constant K_2 increases, and thereby, the slope of the output voltage versus speed relationship increases whereas the slope of the output current versus torque relationship decreases. Therefore, the output voltage E_O of a generator operating at a fixed speed can be varied by varying the field current I_f. This produces the equivalent of a dc source whose output voltage can be controlled by the field current I_f. Figure 2-27 shows the variation of output voltage E_O for a separately-excited dc generator operating at a fixed speed, when the field current I_f is varied over the range from zero to its nominal value.

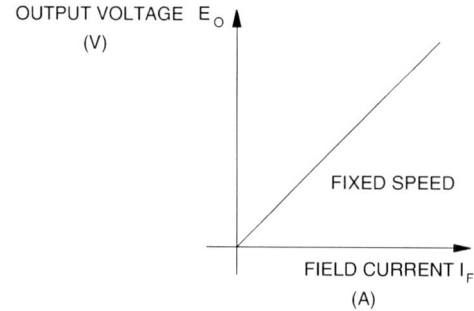

Figure 2-27. E_O versus I_f for a separately-excited dc generator operating at a fixed speed.

The simplified equivalent electric circuit of a separately-excited dc generator is shown in Figure 2-28. It is the same as that for the dc motor, except that the direction of current flow is reversed and voltage E_{CEMF} becomes E_{EMF}, which is the voltage induced across the armature winding as it rotates in the magnetic flux produced by the stator electromagnet. When no load is connected to the dc generator output, the output current I_O is zero and the output voltage E_O equals E_{EMF}.

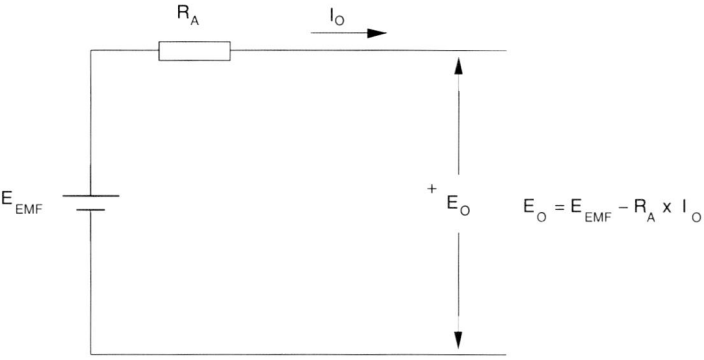

Figure 2-28. Simplified equivalent circuit of a dc generator.

In the first exercise of this unit, you observed that when a fixed armature voltage E_A is applied to a separately-excited dc motor, its speed decreases as the armature current I_A increases. You found that this decrease in speed is due to the armature resistance R_A. Similarly, when the same motor operates as a generator and at a fixed speed, the armature resistance causes the output voltage E_O to decrease with increasing output current as shown in Figure 2-29. The output voltage E_O can be calculated using the following equation:

$$E_O = E_{EMF} - R_A \times I_O$$

where E_O is the dc generator output voltage.
        E_{EMF} is the voltage induced across the armature winding.
        R_A is the armature resistance.
        I_O is the dc generator output current.

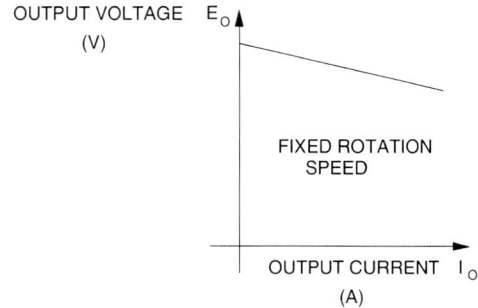

Figure 2-29. Voltage versus current characteristic of a separately-excited dc generator (fixed speed).

Ex. 2-3 – Separately-Excited, Shunt, and Compound DC Generators ◆ *Discussion*

The separately-excited dc generator provides flexible use because its characteristics can be changed by changing the field current. However, a separate dc power source is needed to excite the field electromagnet. This was a disadvantage when the first dc generators were used because dc sources were not commonly available at the time. Therefore, dc generators that operate without a dc power source were designed. These are referred to as self-excited dc generators.

In a self-excited dc generator, the field electromagnet is a shunt winding connected across the generator output (shunt generator) or a combination of a shunt winding connected across the generator output and a series winding connected in series with the generator output (compound generator). The generator output voltage and/or current excite(s) the field electromagnet. The way the field electromagnet is implemented (shunt or compound) determines many of the generator's characteristics.

Self-excitation is possible because of the residual magnetism in the stator pole pieces. As the armature rotates, a small voltage is induced across its winding and a small current flows in the shunt field winding. If this small field current is flowing in the proper direction, the residual magnetism is reinforced which further increases the armature voltage. Thus, a rapid voltage build-up occurs. If the field current flows in the wrong direction, the residual magnetism is reduced and voltage build-up cannot occur. In this case, reversing the connections of the shunt field winding corrects the situation.

In a self-excited dc generator, the output voltage after build-up could be of the opposite polarity to that required. This can be corrected by stopping the generator and setting the polarity of the residual magnetism. To set the residual magnetism, a dc source is connected to the shunt field winding to force nominal current flow in the proper direction. Interrupting the current suddenly sets the polarity of the magnetic poles in the shunt field winding. When the generator is started once again, voltage build-up at the proper polarity occurs.

Figure 2-30 is a graph that shows the voltage versus current characteristics of various types of dc generators. As can be seen, the separately-excited dc generator and the shunt generator have very similar characteristics. The difference is that the output voltage of the shunt generator decreases a little more than that of the separately-excited dc generator as the output current increases. In both cases, the output voltage decreases because the voltage drop across the armature resistor increases as the output current increases. In the shunt generator, the voltage across the shunt field winding, and thereby, the field current, decreases as the output voltage decreases. This causes the output voltage to decrease a little more.

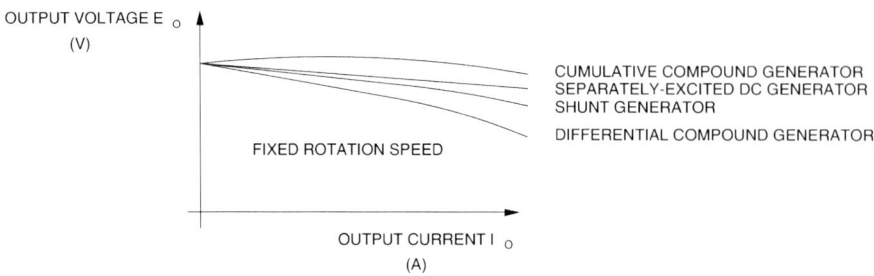

Figure 2-30. Voltage versus current characteristics of various dc generators.

It is possible to compensate for the variation in output voltage by automatically changing the magnetic flux produced by the field electromagnet as the output current varies. The shunt and series field windings of a compound generator can be connected so that the magnetic flux increases when the output current increases. Thus, the output voltage remains fairly constant and changes very little as the output current increases as shown in Figure 2-30. This type of connection results in a cumulative compound generator because the magnetic fluxes created by the two field windings add together in a cumulative manner. For other applications where the output voltage must decrease rapidly when the output current increases, the shunt and series windings can be connected so the magnetic fluxes subtract from each other, resulting in a differential compound generator.

Procedure summary

In the first part of the exercise, you will set up the equipment in the Workstation, connect the equipment as shown in Figure 2-31, and make the appropriate settings on the equipment.

In the second part of the exercise, you will set the field current of the separately-excited dc generator to the same value as that used in Exercise 2-1. You will measure data and plot a graph of the output voltage E_O versus speed n when no electrical load is connected to the generator output. You will calculate the slope of the voltage versus speed relationship and compare it to constant K_1 determined in Exercise 2-1 when the DC Motor/Generator was operating as a separately-excited dc motor.

In the third part of the exercise, you will connect an electrical load to the generator output (setup shown in Figure 2-32), measure data, and plot a graph of the output current I_O versus the applied torque T when the separately-excited dc generator rotates at a fixed speed. You will calculate the slope of the current versus torque relationship and compare it to constant K_2 determined in Exercise 2-1 when the DC Motor/Generator was operating as a separately-excited dc motor.

In the fourth part of the exercise, you will vary the field current I_f of the separately-excited dc generator and observe how the output voltage is affected.

In the fifth part of the exercise, you will use the data obtained in the third part of the exercise to plot a graph of the output voltage versus output current when the separately-excited dc generator operates at a fixed speed.

EQUIPMENT REQUIRED

Refer to the Equipment Utilization Chart in Appendix C to obtain the list of equipment required for this exercise.

PROCEDURE

High voltages are present in this laboratory exercise. Do not make or modify any banana jack connections with the power on unless otherwise specified.

Setting up the equipment

1. Install the equipment required in the EMS workstation.

    If you are performing the exercise using the EMS system, ensure that the brushes of the DC Motor/Generator are adjusted to the neutral point. To do so, connect an ac power source (terminals 4 and N of the Power Supply) to the armature of the DC Motor/Generator (terminals 1 and 2) through CURRENT INPUT I1 of the data acquisition module. Connect the shunt winding of the DC Motor/Generator (terminals 5 and 6) to VOLTAGE INPUT E1 of the data acquisition module. Start the Metering application and open setup configuration file ACMOTOR1.DAI. Turn the Power Supply on and set the voltage control knob so that an ac current (indicated by meter I line 1) equal to half the nominal value of the armature current flows in the armature of the DC Motor/Generator. Adjust the brush adjustment lever on the DC Motor/Generator so that the voltage across the shunt winding (indicated by meter E line 1) is minimum. Turn the Power Supply off, exit the Metering application, and disconnect all leads and cable.

   Mechanically couple the prime mover/dynamometer module to the DC Motor/Generator using a timing belt.

2. On the Power Supply, make sure the main power switch is set to the O (off) position, and the voltage control knob is turned fully counterclockwise. Ensure the Power Supply is connected to a three-phase power source.

    If you are using the Four-Quadrant Dynamometer/Power Supply, Model 8960-2, connect its POWER INPUT to a wall receptacle.

3. Ensure that the data acquisition module is connected to a USB port of the computer.

   Connect the POWER INPUT of the data acquisition module to the 24 V - AC output of the Power Supply.

    If you are using the Prime Mover/Dynamometer, Model 8960-1, connect its LOW POWER INPUT to the 24 V - AC output of the Power Supply.

   On the Power Supply, set the 24 V - AC power switch to the I (on) position.

    If you are using the Four-Quadrant Dynamometer/Power Supply, Model 8960-2, turn it on by setting its POWER INPUT switch to the I (on) position. Press and hold the FUNCTION button 3 seconds to have uncorrected torque values on the display of the Four-Quadrant Dynamometer/Power Supply. The indication "NC" appears next to the function name on the display to indicate that the torque values are uncorrected.

4. Start the Data Acquisition software (LVDAC or LVDAM). Open setup configuration file DCMOTOR1.DAI.

 If you are using LVSIM-EMS in LVVL, you must use the IMPORT option in the File menu to open the configuration file.

In the Metering window, select layout 1. Make sure that the continuous refresh mode is selected.

5. Set up the separately-excited dc generator circuit shown in Figure 2-31. Notice that no electrical load is connected to the generator output.

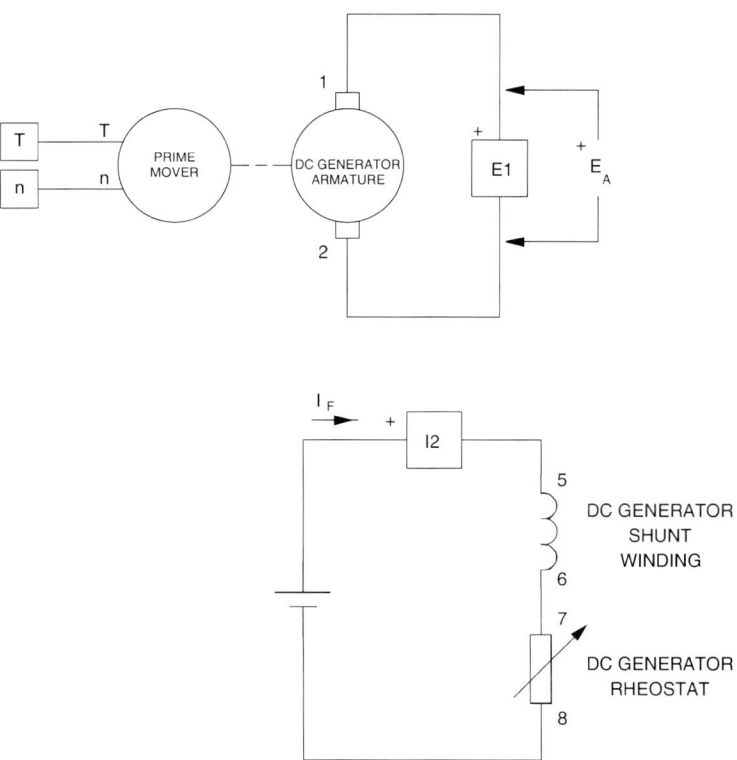

Figure 2-31. Separately-excited dc generator coupled to a prime mover (no electrical load).

6. Set the Four-Quadrant Dynamometer/Power Supply or the Prime Mover/Dynamometer to operate as a clockwise prime mover. To do this, refer to Exercise 1-1 or Exercise 1-2 if necessary.

 If you are performing the exercise using LVSIM®-EMS, you can zoom in on the Prime Mover/Dynamometer before setting the controls in order to see additional front panel markings related to these controls.

Output voltage versus speed characteristic of a separately-excited dc generator

7. Turn the Power Supply on and start the prime mover.

   On the DC Motor/Generator, set the FIELD RHEOSTAT so that the field current I_f indicated by meter I field (I_f) in the Metering window is equal to the value given in Table 2-4.

   Table 2-4. Field current of the separately-excited dc generator.

Local ac power network		I_f (mA)
Voltage (V)	Frequency (Hz)	
120	60	300
220	50	190
220	60	190
240	50	210

8. In the Metering window, make sure that the torque correction function of the Torque meter is enabled. The Torque meter now indicates the torque produced by the dc generator. This torque opposes to rotation. It is equal in magnitude to the torque applied to the dc generator's shaft but of opposite polarity. This explains why the torque indicated by the Torque meter is negative.

   Record the dc generator output voltage E_O, field current I_f, torque T, and speed n in the Data Table. These parameters are indicated by meters E arm. (E_A), I field (I_f), Torque, and Speed, respectively.

   Increase the prime mover speed (indicated by the Speed meter) by 150 r/min increments up to 1500 r/min (150, 300, 450 r/min etc.). For each speed setting, record the data in the Data Table.

9. When all data has been recorded, set the prime mover speed to 0, then turn the Power Supply off.

   In the Data Table window, confirm that the data has been stored, entitle the Data Table as DT231, and print the Data Table.

    Refer to the user guide dealing with the computer-based instruments for EMS to know how to edit, entitle, and print a Data Table.

10. In the Graph window, make the appropriate settings to obtain a graph of the separately-excited dc generator output voltage [obtained from meter E arm. (E_A)] as a function of the speed n (obtained from the Speed meter). Entitle the graph as G231, name the x-axis as Separately-excited dc generator speed, name the y-axis as Separately-excited dc generator output voltage, and print the graph.

Refer to the user guide dealing with the computer-based instruments for EMS to know how to use the Graph window of the Metering application to obtain a graph, entitle a graph, name the axes of a graph, and print a graph.

Does this graph confirm that the separately-excited dc generator is equivalent to a linear speed-to-voltage converter, with higher speed producing greater output voltage?

❏ Yes   ❏ No

11. Use the two end points to calculate the slope of the relationship obtained in graph G231. The values of these points are indicated in Data Table DT231.

$$\text{Slope} = \frac{E_2 - E_1}{n_2 - n_1} = \frac{-}{-} = \frac{}{} \frac{V}{r/min}$$

Compare the slope of the output voltage versus speed relationship to constant K_1 obtained in Exercise 2-1.

___

___

___

In the Data Table window, clear the recorded data.

Output current versus torque characteristic of a separately-excited dc generator

12. Modify the connections to connect a resistive load (R_1) across the separately-excited dc generator output, as shown in the circuit of Figure 2-32. Connect the three resistor sections on the Resistive Load module in parallel to implement resistor R_1.

The values of various components (resistors, inductors, capacitors, etc.) in the circuits used in this manual depend on your local ac power network voltage. Whenever necessary, a table in the circuit diagram indicates the value of each component for ac power network voltages (line voltages) of 120 V, 220 V, and 240 V. Use the component values corresponding to your local ac power network voltage.

Ex. 2-3 – Separately-Excited, Shunt, and Compound DC Generators ◆ *Procedure*

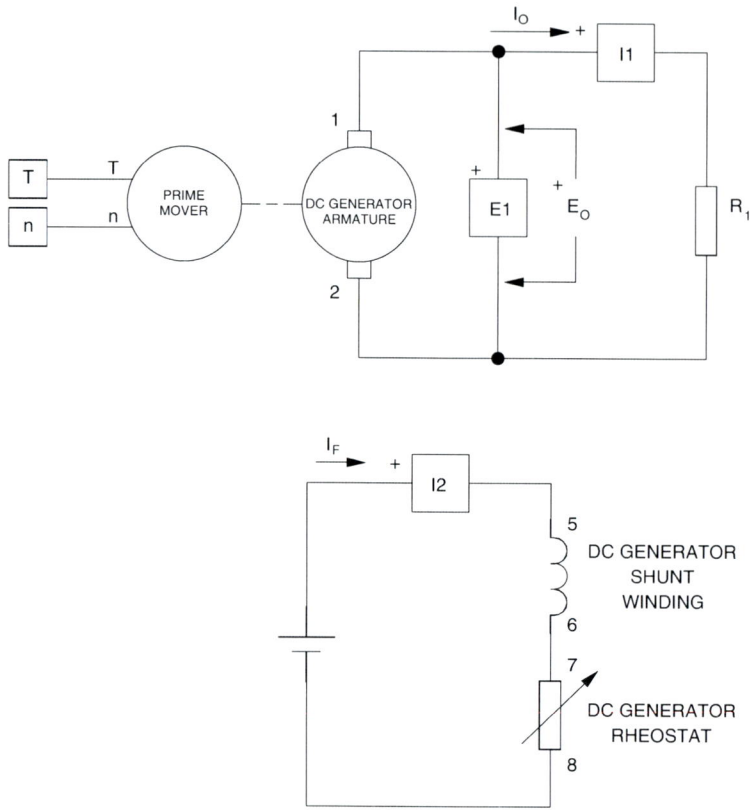

Local ac power network		R_1 (Ω)
Voltage (V)	Frequency (Hz)	
120	60	∞
220	50	∞
220	60	∞
240	50	∞

Figure 2-32. Separately-excited dc generator coupled to a prime mover (with an electrical load).

13. Turn the Power Supply on and start the prime mover.

On the DC Motor/Generator, slightly readjust the FIELD RHEOSTAT so that the field current I_f indicated by meter I field (I_f) still equals the value given in Table 2-4 (if necessary).

Set the prime mover speed so that it is equal to the nominal speed of the DC Motor/Generator.

Ex. 2-3 – Separately-Excited, Shunt, and Compound DC Generators ♦ *Procedure*

14. Record the dc generator output voltage E_O, output current I_O, field current I_f, torque, and speed in the Data Table. These parameters are indicated by meters E arm. (E_A), I arm. (I_A), I field (I_f), Torque, and Speed, respectively.

Modify the settings on the Resistive Load module so that the resistance of resistor R_1 decreases by steps as indicated in Table 2-5. You can refer to Appendix B of this manual to know how to obtain the various resistance values given in Table 2-5. For each resistance setting, readjust the speed setting (if necessary) so that the prime mover speed remains equal to the nominal speed of the DC Motor/Generator and then record the data in the Data Table.

The dc generator output voltage may exceed the rated voltage of the Resistive Load module while performing this manipulation. It is, therefore, suggested to complete the manipulation within a time interval of 5 minutes or less.

Table 2-5. Decreasing R_1 to load the dc generator.

Local ac power network		R_1 (Ω)	R_1 (Ω)	R_1 (Ω)	R_1 (Ω)	R_1 (Ω)	R_1 (Ω)	R_1 (Ω)	R_1 (Ω)
Voltage (V)	Frequency (Hz)								
120	60	1200	600	300	171	120	86	71	57
220	50	4400	2200	1100	629	440	314	259	210
220	60	4400	2200	1100	629	440	314	259	210
240	50	4800	2400	1200	686	480	343	282	229

15. When all data has been recorded, set the prime mover speed to 0 then turn the Power Supply off.

In the Data Table window, confirm that the data has been stored. Reverse the polarity of the torque values indicated in the Torque column to obtain the torque applied to the dc generator's shaft. Entitle the Data Table as DT232, and print the Data Table.

16. In the Graph window, make the appropriate settings to obtain a graph of the dc generator output current [obtained from meter I arm.(I_A)] as a function of the torque T (obtained from the Torque meter). Entitle the graph as G232, name the x-axis as Torque applied to the separately-excited dc generator, name the y-axis as Separately-excited dc generator output current, and print the graph.

The torque is not zero when the output current is zero because some torque is required to overcome opposition to rotation due to friction in the dc generator.

Does this graph confirm that the separately-excited dc generator is equivalent to a linear torque-to-current converter, with higher torque producing greater output current?

❏ Yes ❏ No

17. Use the two end points to calculate the slope of the relationship obtained in graph G232. The values of these points are indicated in Data Table DT232.

$$\text{Slope} = \frac{I_2 - I_1}{T_2 - T_1} = \frac{-}{-} = \frac{\quad}{\quad} \frac{A}{N \cdot m \ (lbf \cdot in)}$$

Compare the slope of the output current versus torque relationship to constant K_2 obtained in Exercise 2-1.

_____

_____

_____

Output voltage versus field current of a separately-excited dc generator

18. On the Resistive Load module, set the resistance of resistor R_1 to the value given in Table 2-6.

Table 2-6. Resistance of resistor R_1.

| Local ac power network || R_1 |
Voltage (V)	Frequency (Hz)	(Ω)
120	60	171
220	50	629
220	60	629
240	50	686

Turn the Power Supply on.

On the DC Motor/Generator, slightly readjust the FIELD RHEOSTAT so that the field current I_f indicated by meter I field (I_f) still equals the value given in Table 2-4 (if necessary).

Set the prime mover speed so that it is equal to the nominal speed of the DC Motor/Generator.

Note below the dc generator output voltage E_O and field current I_f indicated by meters E arm. (E_A) and I field (I_f), respectively:

$E_O = $ _____ V

$I_f = $ _____ A

Ex. 2-3 – Separately-Excited, Shunt, and Compound DC Generators ◆ *Procedure*

19. On the DC Motor/Generator, slowly turn the FIELD RHEOSTAT knob fully clockwise so that the field current I_f increases. While doing this, observe the output voltage E_O indicated by meter E arm. (E_A).

Note the output voltage E_O and field current I_f in the following blank spaces:

$E_O = $ _____ V

$I_f = $ _____ A

On the DC Motor/Generator, set the FIELD RHEOSTAT to the mid position.

Describe what happens to the output voltage E_O when the field current I_f is increased.

_____

_____

20. On the DC Motor/Generator, slowly turn the FIELD RHEOSTAT knob fully counterclockwise so that the field current I_f decreases. While doing this, observe the output voltage E_O indicated by meter E arm. (E_A).

Note the output voltage E_O and field current I_f in the following blank spaces:

$E_O = $ _____ V

$I_f = $ _____ A

Describe what happens to the output voltage E_O when the field current I_f is decreased.

_____

_____

Is a separately-excited dc generator equivalent to a dc power source with variable output voltage?

❏ Yes   ❏ No

Set the prime mover speed to 0, then turn the Power Supply off.

Voltage versus current characteristic of a separately-excited dc generator operating at a fixed speed

21. In the Graph window, make the appropriate settings to obtain a graph of the separately-excited dc generator output voltage E_O [(obtained from meter E arm. (E_A)] as a function of the separately-excited dc generator output current I_O [obtained from meter I arm. (I_A)] using the data recorded previously in the Data Table (DT232). Entitle the graph as G232-1, name the x-axis as Separately-excited dc generator output current, name the y-axis as Separately-excited dc generator output voltage, and print the graph.

Describe how the output voltage E_O varies as the output current I_O increases.

_____

_____

22. On the Power Supply, set the 24 V - AC power switch to the O (off) position.

 If you are using the Four-Quadrant Dynamometer/Power Supply, Model 8960-2, turn it off by setting its POWER INPUT switch to the O (off position).

Remove all leads and cables.

Additional experiments

Voltage versus current characteristic of a shunt generator operating at a fixed speed

You can obtain the output voltage versus output current characteristic of a shunt generator and compare it to that obtained for the separately-excited dc generator. To do so, make sure the Power Supply is turned off and connect terminals 8 and N of the Power Supply to terminals 5 and 6 of the DC Motor/Generator, respectively. Turn the Power Supply on then turn it off. This sets the polarity of the residual magnetism. Set up the shunt generator circuit shown in Figure 2-33.

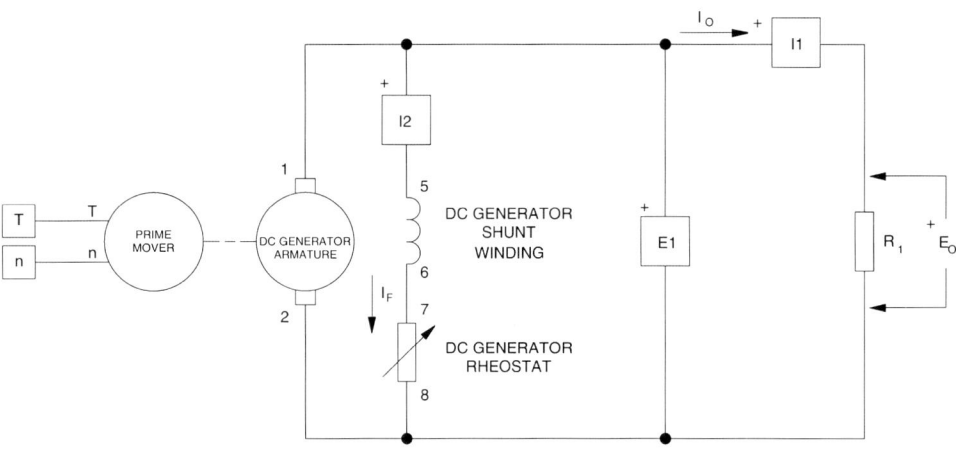

Local ac power network		R_1 (Ω)
Voltage (V)	Frequency (Hz)	
120	60	∞
220	50	∞
220	60	∞
240	50	∞

Figure 2-33. Shunt generator coupled to a prime mover (with an electrical load).

Make sure the Four-Quadrant Dynamometer/Power Supply or the Prime Mover/Dynamometer is set to operate as a clockwise prime mover. (If you are using the Four-Quadrant Power Supply/Dynamometer, make sure to press and hold the FUNCTION button 3 seconds to have uncorrected values on this module's display). Set the FIELD RHEOSTAT on the DC Motor/Generator to the mid position. Turn the Power Supply on. Adjust the prime mover speed so that it is equal to the nominal speed of the DC Motor/Generator. Slightly turn the FIELD RHEOSTAT on the DC Motor/Generator so that the field current I_f is equal to the value indicated in Table 2-4. Clear the data recorded in the Data Table. In the Metering window, make sure that the torque correction function of the Torque meter is enabled. Refer to steps 14, 15, and 21 of this exercise to record the necessary data and obtain the graph. Entitle the Data Table and graph as DT233 and G233, respectively. Compare the output voltage versus output current characteristic of the shunt generator (graph G233) to that of the separately-excited dc generator (graph G232-1).

The output voltage of the shunt generator decreases rapidly as the output current increases because the armature resistance of the DC Motor/Generator is quite large. This is also due to another phenomenon which is called armature reaction. This phenomenon will be studied in the next unit of this manual.

Voltage versus current characteristic of a cumulative compound generator operating at a fixed speed

You can obtain the output voltage versus output current characteristic of a cumulative compound generator and compare it to that obtained for the separately-excited dc generator. To do this, carry out the same manipulations as those used to obtain the voltage versus current characteristic of the shunt generator using the circuit of a cumulative compound generator shown in Figure 2-34. Entitle the Data Table and graph as DT234 and G234, respectively. Compare the output voltage versus output current characteristic of the cumulative compound generator (graph G234) to those of the separately-excited dc generator (graph G232-1) and shunt generator (graph G233).

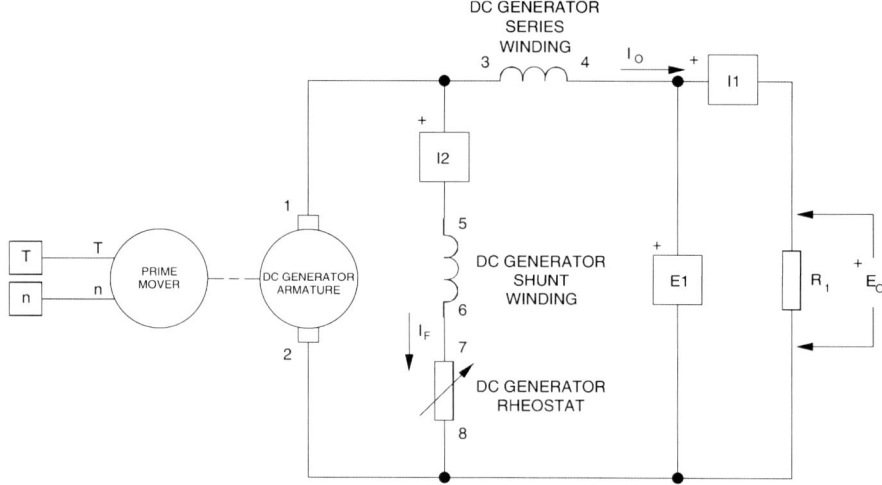

Local ac power network		R_1 (Ω)
Voltage (V)	Frequency (Hz)	
120	60	∞
220	50	∞
220	60	∞
240	50	∞

Figure 2-34. Cumulative compound generator coupled to a prime mover (with an electrical load).

Voltage versus current characteristic of a differential compound generator operating at a fixed speed

You can obtain the output voltage versus output current characteristic of a differential compound generator and compare it to that obtained for the separately-excited dc generator. To do so, carry out the same manipulations as those used to obtain the voltage versus current characteristic of the shunt generator using the circuit of a differential compound generator shown in Figure 2-35. Entitle the Data Table and graph as DT235 and G235, respectively. Compare the voltage versus current characteristic of the differential compound generator (graph G235) to those obtained with the other types of dc generators (graphs G232-1, G233, and G234).

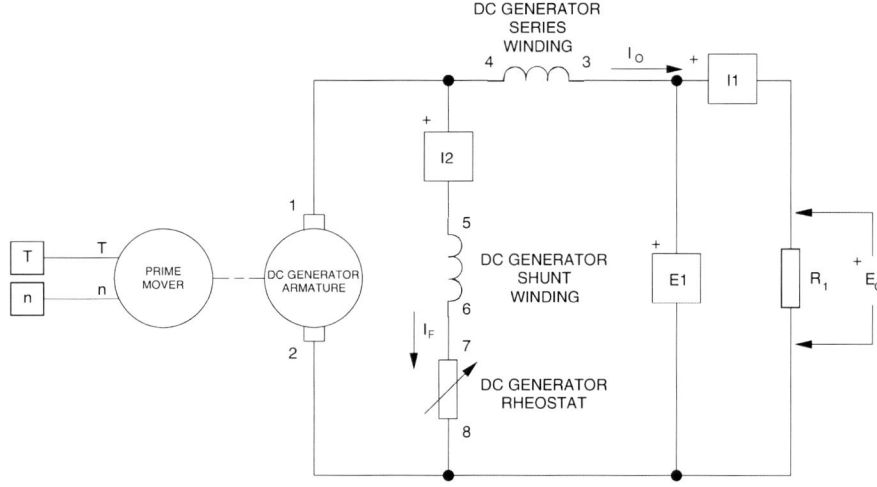

Local ac power network		R_1 (Ω)
Voltage (V)	Frequency (Hz)	
120	60	∞
220	50	∞
220	60	∞
240	50	∞

Figure 2-35. Differential compound generator coupled to a prime mover (with an electrical load).

Ex. 2-3 – Separately-Excited, Shunt, and Compound DC Generators ◆ *Conclusion*

CONCLUSION

In this exercise, you plotted graphs of the main operating characteristics of a separately-excited dc generator. You observed that the output voltage increases linearly with speed. You also observed that the output current increases linearly with the input torque. You found that the slope of the output voltage versus speed characteristic is equal to the reciprocal of constant K_1, and that the slope of the output current versus torque characteristic is equal to the reciprocal of constant K_2. You saw that constants K_1 and K_2 can be changed by changing the field current and that this allows the output voltage to be changed. You observed that the output voltage decreases as the output current increases.

If you have performed the additional experiments, you plotted graphs of the voltage versus current characteristics for shunt, cumulative compound, and differential compound generators. You compared the various voltage versus current characteristics obtained in the exercise. You observed that the output voltage of the shunt generator decreases more rapidly than that of the separately-excited dc generator when the output current increases. You found that the output voltage of a cumulative compound generator varies little as the output current varies. Finally, you saw that the output voltage of a differential compound generator decreases more rapidly than that of the separately-excited and shunt generators when the output current increases.

REVIEW QUESTIONS

1. What effect does decreasing the field current have on the output voltage of a separately-excited dc generator operating at fixed speed?

    a. The output voltage increases.
    b. The output voltage decreases.
    c. The output voltage oscillates around its original value.
    d. The value of the field current has no effect on the output voltage.

2. What effect does increasing the output current have on the input torque of a separately-excited dc generator?

    a. The torque increases.
    b. The torque decreases.
    c. The torque oscillates around its original value.
    d. The value of output current has no effect on the torque.

3. What is the main characteristic of a cumulative compound generator?

    a. The output voltage becomes unstable when the output current decreases.
    b. The output voltage decreases when the output current increases.
    c. The output voltage increases when the output current increases.
    d. The output voltage varies little when the output current varies.

4. What is the main characteristic of a differential compound generator?

   a. The output voltage becomes unstable when the output current decreases.
   b. The output voltage decreases fairly rapidly when the output load current increases.
   c. The output voltage increases when the output current increases.
   d. The output voltage is made independent of the output current.

5. What happens when the field current of a separately-excited dc generator is increased and the speed is maintained constant?

   a. The output current decreases.
   b. The output voltage increases.
   c. The output voltage decreases.
   d. The output voltage is independent of the field current.

Unit Test

1. The rotor, or armature, of a dc motor consists of

   a. an iron cylinder and windings.
   b. an iron cylinder, windings, and brushes.
   c. an iron cylinder, windings, and a commutator.
   d. an iron cylinder, windings, a commutator, and a dc source.

2. The basic principle of operation of a dc motor is

   a. the creation of an electromagnet.
   b. the creation of a rotating electromagnet inside the armature.
   c. the creation of a fixed electromagnet inside the armature.
   d. the creation of a rotating electromagnet at the stator.

3. The speed n of a separately-excited dc motor is equal to

   a. $K_2 \times E_{CEMF}$
   b. $K_1 \times I_A$
   c. $K_1 \times E_{CEMF} \times I_A$
   d. $K_1 \times E_{CEMF}$

4. The armature resistance R_A and constants K_1 and K_2 of a separately-excited dc motor are 0.2 Ω, 8 r/min/V, and 0.8 N·m/A (7.08 lbf·in/A), respectively. What are the speed n and torque T of this motor, when the armature voltage E_A and current I_A are 300 V and 100 A?

   a. n = 2400 r/min, T = 80 N·m (708 lbf·in)
   b. n = 2240 r/min, T = 800 N·m (7080 lbf·in)
   c. n = 2240 r/min, T = 80 N·m (708 lbf·in)
   d. n = 2400 r/min, T = 240 N·m (2124 lbf·in)

5. The field current of a separately-excited dc motor operating with a fixed armature voltage and a fixed mechanical load is changed. This causes the speed to increase. The field current has been

   a. decreased.
   b. increased.
   c. This is not possible because the speed is independent of the field current.
   d. None of the above.

6. When the field current of a separately-excited dc motor is increased,

   a. constants K_1 and K_2 decrease.
   b. constant K_1 decreases and constant K_2 increases.
   c. constant K_1 increases and constant K_2 decreases.
   d. constants K_1 and K_2 increase.

7. The speed of a separately-excited dc motor

   a. increases linearly as the motor torque increases.
   b. decreases linearly as the motor torque increases.
   c. is constant as the motor torque increases.
   d. decreases rapidly and non-linearly as the motor torque increases.

8. In a series motor, the field electromagnet consists of

   a. a winding connected in parallel with the armature.
   b. a winding connected in parallel with the armature and a second winding connected in series with the armature.
   c. a winding connected in series with the armature.
   d. a winding connected in series with a separate dc power source.

9. The voltage induced in a separately-excited dc generator (E_{EMF}) that rotates at a fixed speed of 1600 r/min is 600 V. This causes a current of 400 A to flow in the electrical load connected across the dc generator. What is the output voltage E_O of the generator when its armature resistance is 0.15 Ω?

   a. $E_O = 360$ V
   b. $E_O = 540$ V
   c. $E_O = 600$ V
   d. $E_O = 200$ V

10. The output voltage E_O of a cumulative compound generator

    a. increases linearly as the output current I_O increases.
    b. decreases linearly as the output current I_O increases.
    c. varies little as the output current I_O increases.
    d. decreases rapidly and non-linearly as the output current I_O increases.

Unit 3

Special Characteristics of DC Motors

UNIT OBJECTIVE

When you have completed this unit, you will be able to demonstrate and explain some of the special operating characteristics of dc motors.

DISCUSSION OF FUNDAMENTALS

In Unit 2 of this manual, you observed the main operating characteristics of a separately-excited dc motor, which can be considered as a linear voltage-to-speed converter and a linear current-to-torque converter. You also observed that the dc motor is a reversible converter capable of converting electrical power to mechanical power, and vice-versa.

However, the operation of a dc motor is no longer linear when either the field or armature current exceeds its nominal value. When the field current is too high, the phenomenon of saturation in the iron of the dc machine occurs. Consequently, the flux of the fixed magnetic field in the dc machine no longer increases proportionally to the field current. When the armature current is too high, a phenomenon called armature reaction occurs. Armature reaction causes the flux of the fixed magnetic field in the dc machine to be modified, and thereby, changes the characteristic of torque versus armature current. It also causes a reduction in the induced voltage (E_{CEMF} or E_{EMF} depending on whether the dc machine operates as a motor or a generator).

In Unit 2 of this manual, the operation of shunt and series dc motors connected to a dc power source has been observed. These two motors can also operate from ac power but their performance is poor. In this unit, you will observe that the addition of a special compensating winding allows acceptable performance to be obtained from a series motor operating from an ac power source. This type of series motor is called a universal motor.

Exercise 3-1

Armature Reaction and Saturation Effect

EXERCISE OBJECTIVE

When you have completed this exercise, you will be able to demonstrate some of the effects of armature reaction and saturation in dc machines using the DC Motor/Generator module.

DISCUSSION

Armature reaction

Previously, you saw that the speed of a dc motor or generator is proportional to the armature voltage E_A and the torque is proportional to the armature current I_A. However, these two relationships no longer apply when the armature current I_A considerably increases and exceeds its nominal value. This is because the magnetic field produced by the armature starts to negatively affect the magnetic field produced by the field electromagnet. The effect of armature reaction on the output voltage of a dc generator is illustrated in Figure 3-1.

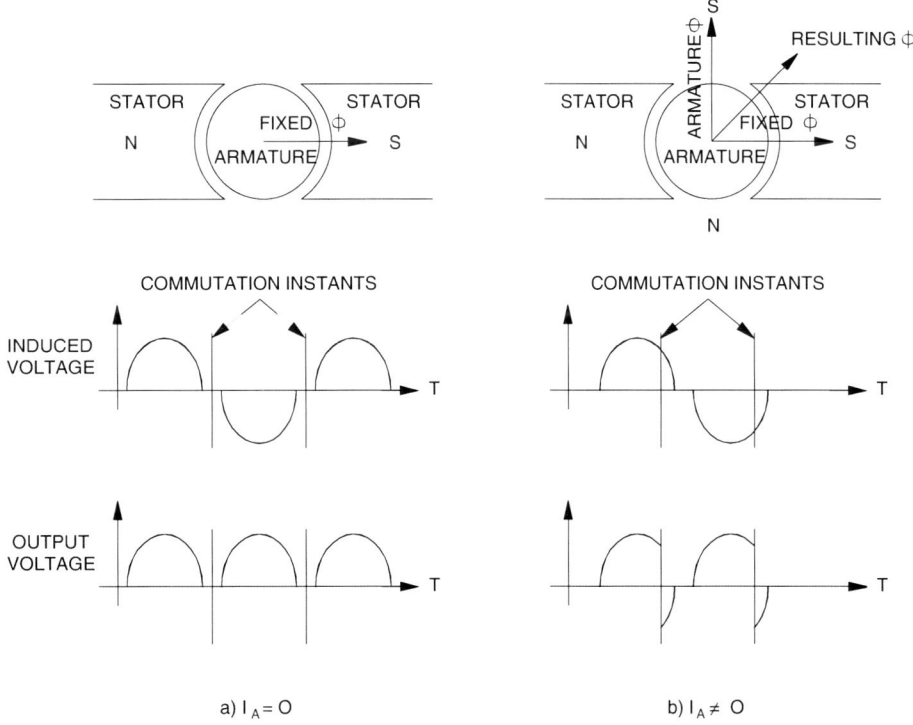

Figure 3-1. Effect of the armature reaction on the output voltage.

Ex. 3-1 – Armature Reaction and Saturation Effect ♦ *Discussion*

When the armature current I_A equals zero, the flux φ in the dc generator is horizontal, the commutator perfectly rectifies the voltage induced in the armature winding, and the dc generator output voltage is maximum, as shown in Figure 3-1a. However, when the armature current I_A does not equal zero, the magnetic fields produced by the armature and the field electromagnet add vectorially. The magnetic flux resulting from the interaction of both magnetic fields is no longer horizontal as shown in Figure 3-1b, and the induced voltage is delayed. Since the instants of commutation have not changed, the average value of the rectified voltage (output voltage) is reduced. Along with producing a lower output voltage, commutation occurs at instants when the induced voltage is not zero, and thus, causes sparking at the brushes and commutator. This increases wear on the brushes and commutator. Another problem created by armature reaction is a decrease in the magnetic torque when the armature current I_A increases.

Figure 3-2a shows the effect of armature reaction on the output voltage versus output current relationship of a separately-excited dc generator. The dotted line is the voltage versus current relationship for a theoretical dc generator (without armature reaction, i.e., $E_O = E_{EMF} - R_A \times I_O$). The other curve is the actual voltage versus current of the same generator, including armature reaction. As can be seen, armature reaction causes an additional decrease in the output voltage. This additional decrease becomes higher and higher as the output current increases.

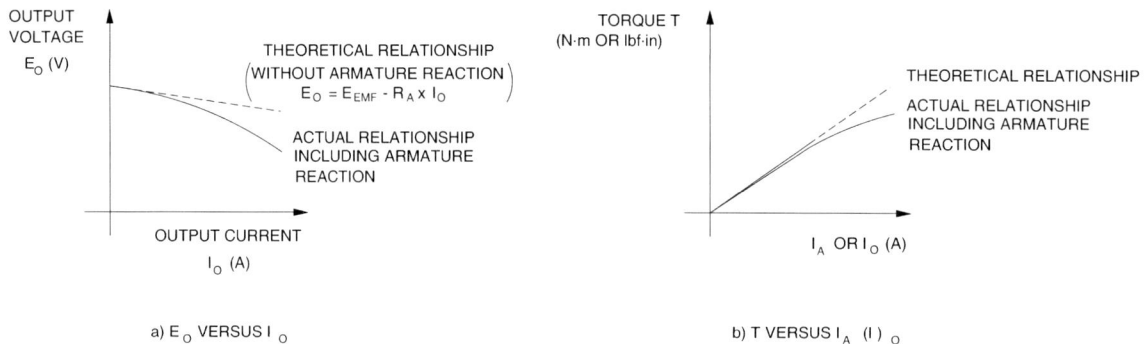

Figure 3-2. Effects of the armature reaction.

Figure 3-2b shows the effect of armature reaction on the torque versus current relationship of a separately-excited dc machine. The dotted line is the theoretical (linear) torque versus current relationship, i.e., without armature reaction. The other curve is the actual relationship including armature reaction. As can be seen, armature reaction causes the torque to cease increasing linearly with current (I_A or I_O depending on whether the dc machine operates as a motor or generator).

The most serious consequence of armature reaction is the increased wear on the brushes and the commutator caused by sparking. For small dc machines, commutation can be improved by shifting the position of the brushes, but this solution only applies to the exact operating point at which they are adjusted. If one wish to change the direction of rotation or operate the dc machine as a generator, the brush position must be readjusted. To improve commutation, large motors include extra windings, called commutating windings, through which

Ex. 3-1 – Armature Reaction and Saturation Effect ◆ *Discussion*

armature current flows. They are physically located so as to produce a magnetic field that causes a weak voltage to be induced in the armature coils being commutated. In this way, proper commutation is ensured independently of the value of the armature current, the direction of rotation, and the machine operation (motor or generator).

Commutation can also be improved by using a permanent-magnet dc motor because it exhibits almost no armature reaction for values of current up to five times greater than the nominal armature current. This is due to the fact that a permanent magnet can create a very powerful magnetic field that is almost completely immune to being affected by another magnetic source. The magnetic field produced by the armature, therefore, has very little effect on the overall magnetic field in the machine.

Another criteria which influences commutation is the inductance L_A of the armature winding. When the armature inductance is too large, commutation is difficult because current flow cannot stop and reverse instantly in inductors having a large inductance. The permanent-magnet dc motor has the particularity of having a small armature inductance which ensures better commutation. For these reasons, the characteristics of permanent-magnet dc motors exceed those of separately-excited, series, and shunt motors. However, it is not possible to build large-size permanent-magnet dc motors.

Saturation effect

As you saw previously, the field current I_f of a dc motor can be varied to modify the operating characteristics. For example, when I_f is decreased, the speed increases even though the armature voltage remains fixed. However, the motor torque developed for a given armature current decreases. As a result, the motor output power remains the same because it is proportional to the product of speed and torque.

Many times, it is desirable to have a motor that produces a maximum value of torque at low speed. To obtain such a motor, the strength of the field electromagnet must be increased (higher field current I_f), as well as the strength of the rotating electromagnet in the armature (higher armature current I_A). However, the armature current must be limited to prevent overheating. Furthermore, the field current must also be limited to prevent saturation. When one starts to increase the field current, constant K_2 increases proportionally. However, once the field current exceeds a certain value, saturation in the iron of the machine starts to occur. As a result, the strength of the field electromagnet no longer increases proportionally to the field current. Figure 3-3 illustrates how the torque produced by a dc motor increases when the field current I_f increases and the armature current I_A remains to a fixed value.

Ex. 3-1 – Armature Reaction and Saturation Effect ♦ *Discussion*

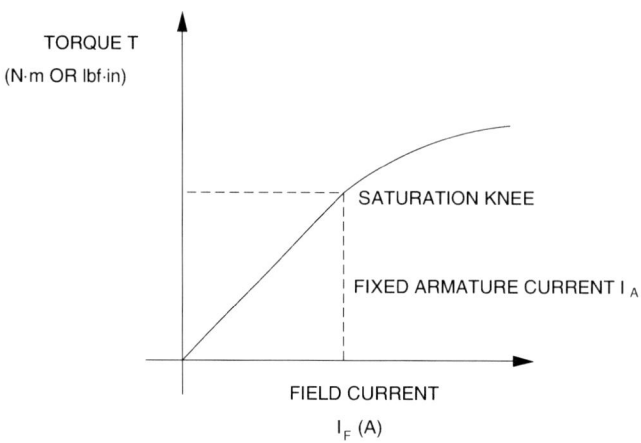

Figure 3-3. Effect of saturation on the torque of a dc motor.

As can be seen, the curve of the torque T versus the field current I_f flattens out for higher values of I_f. The extra increase in torque for additional increases in field current becomes smaller once the saturation knee is exceeded. Higher values of field current also produce more heating in the motor. Usually, the nominal value of the field current is chosen to be just at the beginning of the saturation knee to obtain as much torque as possible with a field current that is as low as possible.

This same characteristic can be visualized using a dc motor as a generator because the stronger the field electromagnet, the higher the induced voltage E_{EMF} at a given speed, and the higher the output voltage E_O. Figure 3-4 shows the relationship between the output voltage E_O and field current I_f for a fixed speed.

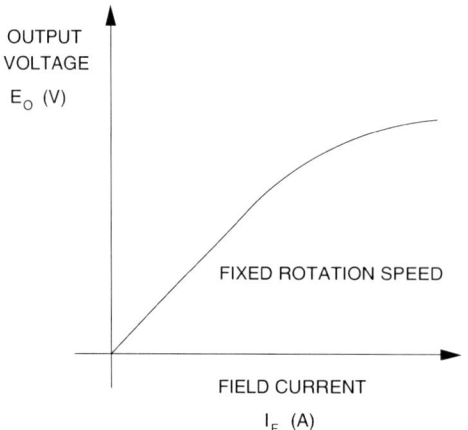

Figure 3-4. Effect of saturation on the output voltage of a dc generator.

Ex. 3-1 – Armature Reaction and Saturation Effect ◆ *Procedure*

Procedure summary

In the first part of this exercise, you will perform calculations with data obtained in Exercises 2-1 and 2-3. You will use the results of these calculations to draw on graph G232-1 the theoretical output voltage versus output current relationship of the separately-excited dc generator used in Exercise 2-3. This will allow you to illustrate the effect of armature reaction on the output voltage of a dc generator.

In the second part of the exercise, you will set up the equipment in the Workstation, connect the equipment as shown in Figure 3-5, and make the appropriate settings on the equipment.

In the third part of the exercise, you will set the field current of the separately-excited dc generator. You will vary the output current of the dc generator from zero to twice its nominal value to obtain the necessary data to plot a graph of the torque applied to the dc generator versus the output current I_O. This will allow you to demonstrate the effect of armature reaction on the torque versus current relationship of a dc machine.

In the fourth part of the exercise, you will connect the equipment as shown in Figure 3-6. You will vary the field current of a separately-excited dc motor from zero to approximately 175% of its nominal value, while maintaining a fixed armature current, to obtain the necessary data to plot a graph of the motor torque versus the field current I_f. This will allow you to demonstrate the effect of saturation in dc machines.

EQUIPMENT REQUIRED

Refer to the Equipment Utilization Chart in Appendix C to obtain the list of equipment required for this exercise.

PROCEDURE

⚠ **WARNING**

High voltages are present in this laboratory exercise. Do not make or modify any banana jack connections with the power on unless otherwise specified.

Effect of the armature reaction on the output voltage of a dc generator

1. Record in the following blank space the armature resistance of the DC Motor/Generator measured in Exercise 2-1.

   $R_A = $ _____ Ω

Ex. 3-1 – Armature Reaction and Saturation Effect ♦ *Procedure*

2. Refer to graph G232-1 obtained in Exercise 2-3. This graph shows the output voltage versus output current relationship of a separately-excited dc generator operating at a fixed speed. Record the no load output voltage (voltage obtained when the dc generator output current $I_O = 0$ A) in the following blank space (this voltage is recorded in Data Table DT232). This voltage is equal to the voltage induced across the armature winding of the dc generator (E_{EMF}).

   $E_{EMF} = $ _____ V

3. Calculate the dc generator output voltage E_O for each of the output currents indicated in Table 3-1 using the following equation:

   $$E_O = E_{EMF} - R_A \times I_O$$

   Table 3-1. DC generator output currents.

Line voltage (V ac)	Output current I_O (A)	Output current I_O (A)	Output current I_O (A)	Output current I_O (A)
120	0.5	1.0	1.5	2.0
220	0.25	0.5	0.75	1.0
240	0.25	0.5	0.75	1.0

   When $I_O = $ _____ A, $E_O = $ _____ V

   When $I_O = $ _____ A, $E_O = $ _____ V

   When $I_O = $ _____ A, $E_O = $ _____ V

   When $I_O = $ _____ A, $E_O = $ _____ V

4. Use the output voltages and currents obtained in the previous step to draw on graph G232-1 the theoretical output voltage versus output current relationship of the separately-excited dc generator.

   Compare the theoretical and actual voltage versus current relationships drawn on graph G232-1. Does this demonstrate that the armature reaction causes an additional decrease in the output voltage as the output current increases?

   ❏ Yes ❏ No

Setting up the equipment

5. Install the equipment required in the EMS workstation.

    If you are performing the exercise using the EMS system, ensure that the brushes of the DC Motor/Generator are adjusted to the neutral point. To do so, connect an ac power source (terminals 4 and N of the Power Supply) to the armature of the DC Motor/Generator (terminals 1 and 2) through CURRENT INPUT I1 of the data acquisition module. Connect the shunt winding of the DC Motor/Generator (terminals 5 and 6) to VOLTAGE INPUT E1 of the data acquisition module. Start the Metering application and open setup configuration file ACMOTOR1.DAI. Turn the Power Supply on and set the voltage control knob so that an ac current (indicated by meter I line 1) equal to half the nominal value of the armature current flows in the armature of the DC Motor/Generator. Adjust the brush adjustment lever on the DC Motor/Generator so that the voltage across the shunt winding (indicated by meter E line 1) is minimum. Turn the Power Supply off, exit the Metering application, and disconnect all leads and cable.

   Mechanically couple the prime mover/dynamometer module to the DC Motor/Generator using a timing belt.

6. On the Power Supply, make sure the main power switch is set to the O (off) position, and the voltage control knob is turned fully counterclockwise. Ensure the Power Supply is connected to a three-phase power source.

    If you are using the Four-Quadrant Dynamometer/Power Supply, Model 8960-2, connect its POWER INPUT to a wall receptacle.

7. Ensure that the data acquisition module is connected to a USB port of the computer.

   Connect the POWER INPUT of the data acquisition module to the 24 V - AC output of the Power Supply.

    If you are using the Prime Mover/Dynamometer, Model 8960-1, connect its LOW POWER INPUT to the 24 V - AC output of the Power Supply.

   On the Power Supply, set the 24 V - AC power switch to the I (on) position.

    If you are using the Four-Quadrant Dynamometer/Power Supply, Model 8960-2, turn it on by setting its POWER INPUT switch to the I (on) position. Press and hold the FUNCTION button 3 seconds to have uncorrected torque values on the display of the Four-Quadrant Dynamometer/Power Supply. The indication "NC" appears next to the function name on the display to indicate that the torque values are uncorrected.

Ex. 3-1 – Armature Reaction and Saturation Effect ♦ *Procedure*

8. Start the Data Acquisition software (LVDAC or LVDAM). Open setup configuration file DCMOTOR1.DAI.

    If you are using LVSIM-EMS in LVVL, you must use the IMPORT option in the File menu to open the configuration file.

   In the Metering window, select layout 1. Make sure that the continuous refresh mode is selected.

9. Set up the separately-excited dc generator circuit shown in Figure 3-5.

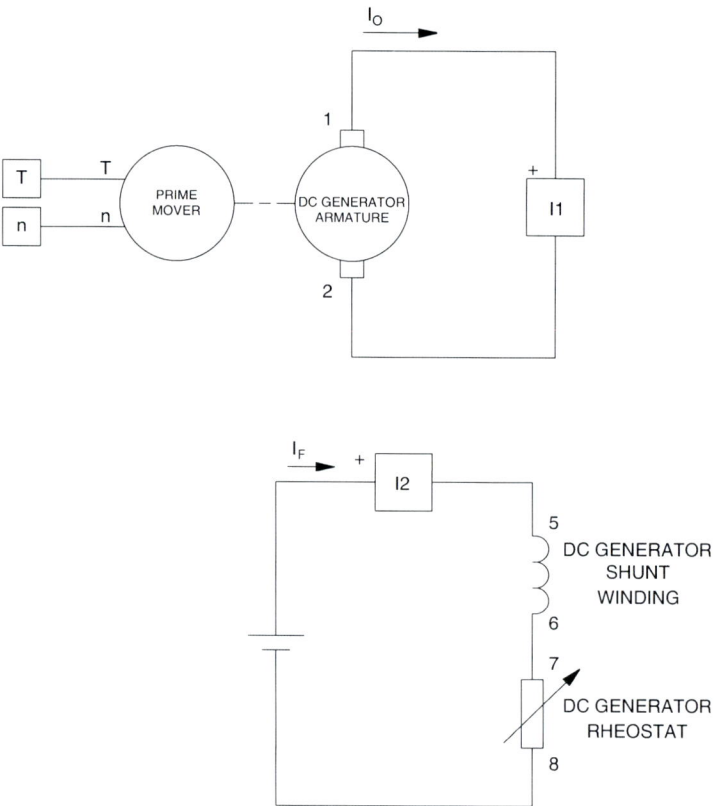

Figure 3-5. Separately-excited dc generator coupled to a prime mover.

10. Set the Four-Quadrant Dynamometer/Power Supply or the Prime Mover/Dynamometer to operate as a clockwise prime mover. To do this, refer to Exercise 1-1 or Exercise 1-2 if necessary.

     If you are performing the exercise using LVSIM®-EMS, you can zoom in on the Prime Mover/Dynamometer before setting the controls in order to see additional front panel markings related to these controls.

Effect of the armature reaction on torque

11. Turn the Power Supply on.

Ex. 3-1 – Armature Reaction and Saturation Effect ◆ *Procedure*

On the DC Motor/Generator, set the FIELD RHEOSTAT so that the field current I_f indicated by meter I field (I_f) in the Metering window is equal to the value given in Table 3-2.

Table 3-2. Field current of the separately-excited dc motor.

Local ac power network		I_f (mA)
Voltage (V)	Frequency (Hz)	
120	60	250
220	50	160
220	60	160
240	50	175

In the Metering window, make sure that the torque correction function of the Torque meter is enabled. The Torque meter now indicates the torque produced by the dc generator. This torque opposes to rotation. It is equal in magnitude to the torque applied to the dc generator's shaft but of opposite polarity.

12. Record the dc generator output current I_O, field current I_f, torque T, and speed n in the Data Table. These parameters are indicated by meters I arm. (I_A), I field (I_f), Torque, and Speed, respectively.

Gradually increase the prime mover speed to increase the generator output current I_O [indicated by meter I arm. (I_A)] to approximately twice the nominal armature current of the DC Motor/Generator in ten steps. For each current setting, record the data in the Data Table.

The rating of any of the machines is indicated in the lower left corner of the module front panel.

The output current exceeds the rated armature current of the DC Motor/Generator while performing this manipulation. It is, therefore, suggested to complete the manipulation within a time interval of 5 minutes or less.

13. When all data has been recorded, set the prime mover speed to 0, then turn the Power Supply off.

In the Data Table window, confirm that the data has been stored. Reverse the polarity of the torque values indicated in the Torque column to obtain the torque applied to the dc generator's shaft. Entitle the Data Table as DT311, and print the Data Table.

Refer to the user guide dealing with the computer-based instruments for EMS to know how to edit, entitle, and print a Data Table.

14. In the Graph window, make the appropriate settings to obtain a graph of the torque applied to the dc generator (obtained from the Torque meter) as a function of the dc generator output current I_O [obtained from meter I arm. (I_A)]. Entitle the graph as G311, name the x-axis as DC generator output current, name the y-axis as Torque applied to the dc generator, and print the graph.

Refer to the user guide dealing with the computer-based instruments for EMS to know how to use the Graph window of the Metering application to obtain a graph, entitle a graph, name the axes of a graph, and print a graph.

Can we say that the variation in torque is linear when the dc generator output current I_O exceeds the nominal armature current of the DC Motor/Generator?

❏ Yes ❏ No

In the Data Table window, clear the recorded data.

Effect of the saturation on torque

15. Modify the connections to obtain the separately-excited dc motor circuit shown in Figure 3-6. Connect the three resistor sections on the Resistive Load module in parallel to implement resistor R_1. Leave the circuit open at points A and B shown in the figure.

    Set the Four-Quadrant Dynamometer/Power Supply or the Prime Mover/Dynamometer to operate as a brake. (If you are using the Four-Quadrant Dynamometer/Power Supply, make sure to press and hold the FUNCTION button 3 seconds to have uncorrected torque values on this module's display). Set the torque control to maximum (fully clockwise position). To do this, refer to Exercise 1-1 or Exercise 1-2 if necessary.

    On the DC Motor/Generator, turn the FIELD RHEOSTAT knob fully counterclockwise

16. In the Metering window, make sure that the torque correction function of the Torque meter is enabled. Turn the Power Supply on, adjust the voltage control knob so that the dc motor armature current indicated by meter I arm. (I_A) is equal to 50% of the nominal value, then record the dc motor armature voltage E_A, armature current I_A, field current I_f, torque T and speed n in the Data Table.

    Turn the voltage control knob fully counterclockwise and turn the Power Supply off.

    Interconnect points A and B shown in the circuit of Figure 3-6.

Ex. 3-1 – Armature Reaction and Saturation Effect ♦ *Procedure*

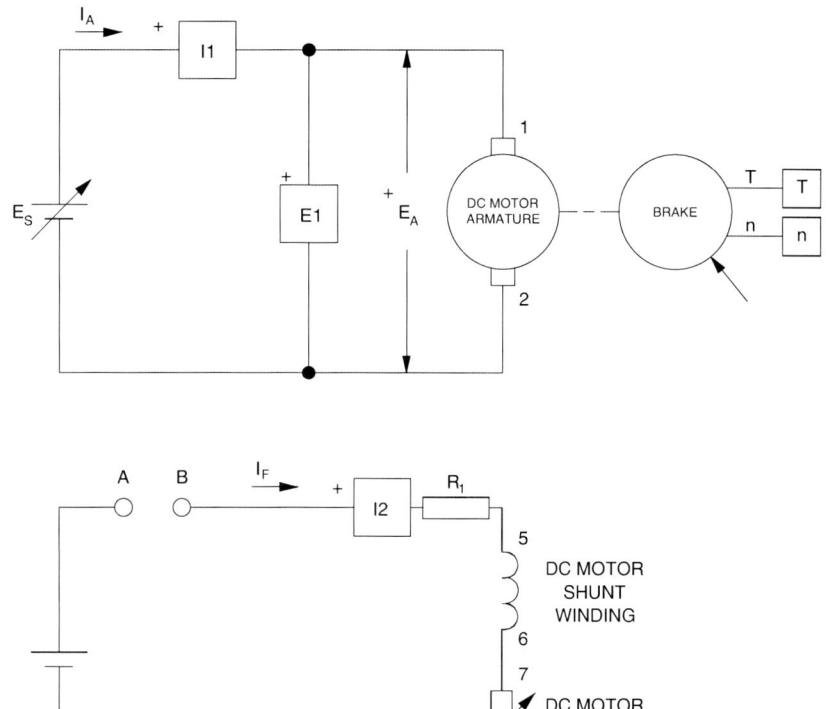

Figure 3-6. Separately-excited dc motor coupled to a brake.

Local ac power network		R_1 (Ω)
Voltage (V)	Frequency (Hz)	
120	60	1200
220	50	4400
220	60	4400
240	50	4800

17. Turn the Power Supply on, adjust the voltage control knob so that the dc motor armature current indicated by meter I arm. (I_A) is equal to 50% of the nominal value, then record the data in the Data Table.

Set the resistance of resistor R_1 (by changing the settings of the toggle switches on the Resistive Load module) and the FIELD RHEOSTAT so that the field current I_f increases by steps as indicated in Table 3-3. For each current setting, readjust the voltage control knob of the Power Supply so that the armature current I_A remains equal to 50% of the nominal value, then record the data in the Data Table.

Resistor R_1 must be short-circuited to obtain the high field currents given in Table 3-3.

Ex. 3-1 – Armature Reaction and Saturation Effect ◆ *Procedure*

Table 3-3. Field currents of the separately-excited dc motor.

Local ac power network		Field current I_f (mA)	Field current I_f (mA)	Field current I_f (mA)	Field current I_f (mA)	Field current I_f (mA)	Field current I_f (mA)	Field current I_f (mA)	Field current I_f (mA)
Voltage (V)	Frequency (Hz)								
120	60	100	150	200	250	300	350	400	450
220	50	65	95	130	160	190	220	255	285
220	60	65	95	130	160	190	220	255	285
240	50	70	105	140	175	210	245	280	315

18. On the DC Motor/Generator, turn the FIELD RHEOSTAT knob fully clockwise, readjust the voltage control knob of the Power Supply so that the armature current remains equal to 50% of the nominal value, then record the data in the Data Table.

    Turn the voltage control knob fully counterclockwise and turn the Power Supply off.

    In the Data Table window, confirm that the data has been stored, entitle the Data Table as DT312, and print the Data Table.

19. In the Graph window, make the appropriate settings to obtain a graph of the dc motor torque (obtained from the Torque meter) as a function of the field current I_f [obtained from meter I field (I_f)]. Entitle the graph as G312, name the x-axis as DC motor field current, name the y-axis as DC motor torque, and print the graph.

    Observe graph G312. How does the dc motor torque vary as the field current increases?

    _____
    _____
    _____

    Briefly explain what happens when the field current exceeds the nominal value.

    _____
    _____
    _____

20. On the Power Supply, set the 24 V - AC power switch to the O (off) position.

 If you are using the Four-Quadrant Dynamometer/Power Supply, Model 8960-2, turn it off by setting its POWER INPUT switch to the O (off position).

Remove all leads and cables.

Additional experiments

Armature inductance of the DC Motor/Generator

You can determine the armature inductance of the DC Motor/Generator. To do so, make sure the Power Supply is turned off and the voltage control knob is turned fully counterclockwise, then set up the circuit shown in Figure 3-7. In the Metering window, open setup configuration file ACMOTOR1.DAI, select meter layout 2, and change the function of the Frequency meter so that it measures reactance X. Turn the Power Supply on and adjust the voltage control knob so that an ac current (indicated by meter I line 1) equal to the nominal armature current of the DC Motor/Generator flows in the armature. Record the armature reactance, X_A, indicated by the reactance meter, turn the Power Supply off, and convert X_A into inductance using the following formula:

$$L_A = \frac{X_A}{2\pi f} \times 1000 = \underline{\hspace{2cm}} = mH$$

 The dc motor should not start rotating when ac power is applied to the armature. If so, slightly readjust the position of the brushes on the DC Motor/Generator so they are exactly at the neutral point. The dc motor should stop rotating, allowing you to perform the armature reactance measurement.

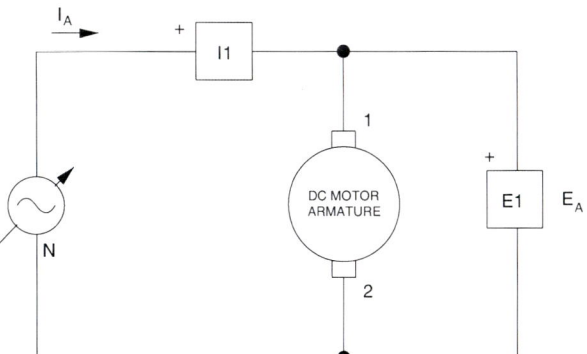

Figure 3-7. Circuit for measuring the armature inductance of the DC Motor/Generator.

Armature inductance of the prime mover (can be measured only when using the Prime Mover/Dynamometer, Model 8960-1)

 This additional experiment cannot be performed when using the Four-Quadrant Dynamometer/Power Supply, Model 8960-2 or 8960-B.

Ex. 3-1 – Armature Reaction and Saturation Effect ♦ *Conclusion*

If you are using the Prime Mover/Dynamometer, Model 8960-1, you can determine the inductance of the armature of the prime mover; this prime mover is a permanent-magnet dc motor. To do this, make sure the Power Supply is turned off and the voltage control knob is turned fully counterclockwise, then set up the circuit shown in Figure 3-8. On the Prime Mover/Dynamometer, set the MODE switch to the PRIME MOVER position. In the Metering window, open setup configuration file ACMOTOR1.DAI, select meter layout 2, and set the function of the Frequency meter so that it measures reactance X. Turn the Power Supply on and adjust the voltage control knob so that a 3-A ac current (indicated by meter I line 1) flows in the prime mover. Record the armature reactance (X_A) indicated by the reactance meter, turn the Power Supply off, and convert X_A into inductance using the following formula:

$$L_A = \frac{X_A}{2\pi f} \times 1000 = \text{_____} = \text{mH}$$

Compare the armature inductance of the prime mover with that of the DC Motor/Generator found in the previous experiment.

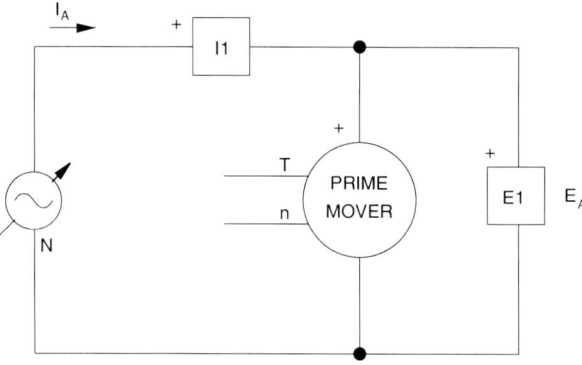

Figure 3-8. Circuit for measuring the armature inductance of the prime mover of the Prime Mover/Dynamometer, Model 8960-1.

Effect of the armature reaction on the torque developed by a dc motor

You can observe the effect which armature reaction has on the torque versus current characteristic of a separately-excited dc motor. To do so, refer to graph G212. This graph shows the torque versus current characteristic of the separately-excited dc motor used in Exercise 2-1. Observe that the torque versus current characteristic is no longer linear for high armature currents.

CONCLUSION

In this exercise, you saw that armature reaction in dc machines causes the output voltage of a generator to decrease rapidly as the armature current increases. You observed that motor torque is also affected in the same manner. You saw that the torque ceases to increase linearly with the field current when the iron in the dc machine begins to saturate.

Ex. 3-1 – Armature Reaction and Saturation Effect ♦ *Review Questions*

If you performed the additional experiments, you found that the armature inductance of the DC Motor/Generator is much higher than that of the prime mover (permanent-magnet dc motor) in the Prime Mover/Dynamometer, Model 8960-1. You observed that armature reaction affects the torque versus current characteristic of a separately-excited dc motor.

REVIEW QUESTIONS

1. What is the most serious consequence of armature reaction in dc machines?

    a. Increased wear on the armature winding.
    b. Increased wear on the field electromagnet and brushes.
    c. Reduced wear on the brushes and commutator.
    d. Increased wear on the brushes and commutator.

2. How does armature reaction affect the output voltage of a dc generator?

    a. The output voltage is lower than it should be.
    b. The output voltage is higher than it should be.
    c. There is no effect.
    d. The output voltage is unstable.

3. Armature reaction in a dc motor causes the torque to be

    a. lower than it should be.
    b. higher than it should be.
    c. highly unstable.
    d. more stable.

4. A permanent-magnet dc motor has better commutation than a conventional dc motor because

    a. its armature inductance is greater.
    b. its armature inductance is smaller.
    c. it has commutating windings.
    d. of the permanent magnets.

5. The brushes on a dc machine having commutating windings

    a. wear out more rapidly.
    b. produce more sparking.
    c. no longer have to be readjusted for different operating points.
    d. Both a and b.

Exercise 3-2

The Universal Motor

EXERCISE OBJECTIVE

When you have completed this exercise, you will be able to demonstrate both ac and dc operation of universal motors using the Universal Motor module.

DISCUSSION

You saw in Unit 2 that the armature winding creates a rotating magnetic field in the rotor of a dc motor. This magnetic field rotates at the same speed as the motor but in the opposite direction. As a result, the poles of the rotor electromagnet remain at a fixed location. Furthermore, the poles of the rotor electromagnet are always at 90° to the poles of the stator magnet or electromagnet (field electromagnet), as was illustrated in Figure 2-5.

However, if either the polarity of the stator electromagnet or that of the rotor electromagnet is reversed, the motor direction of rotation is reversed because the forces of attraction and repulsion between the two magnets are reversed. Figure 3-9 illustrates the different possibilities when the polarities of the armature current I_A and field current I_f are changed. When currents I_A and I_f are of the same polarity, the motor rotates clockwise. Conversely, when currents I_A and I_f are of opposite polarity, it rotates counterclockwise.

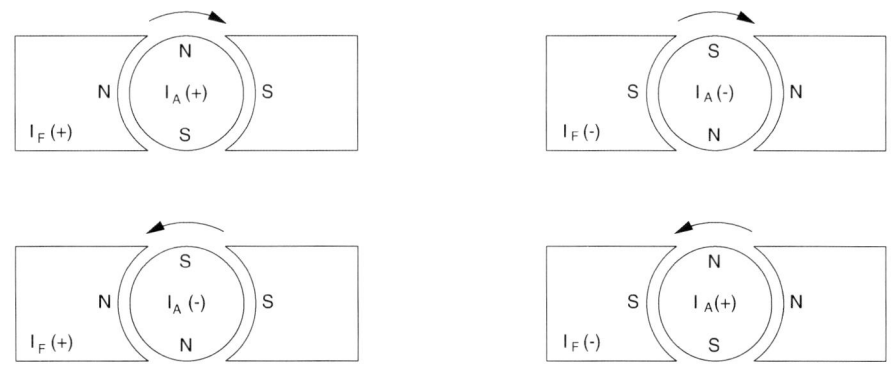

Figure 3-9. Direction of rotation depends on the polarities of the armature and field currents.

When both the armature and the field electromagnet of a dc motor are powered from the same source, which is the case for shunt and series motors, reversing the polarity of the voltage source reverses the polarity of both the armature and field currents. Consequently, the torque does not change direction when the polarity of the voltage applied to the motor changes. Therefore, shunt and series dc motors rotate when connected to an ac power source despite the fact that the source voltage polarity is constantly changing.

However, since motors are made of windings and iron, they always have inductance associated with their windings. For example, the field winding of a shunt motor usually has a large inductance value because it consists of many turns of wire. This makes it difficult for alternating current to flow in the winding because a large inductance means a high impedance. For this reason, it is almost impossible to obtain satisfactory performance from a shunt motor connected to an ac power source.

A series motor has a field winding that consists of only a few turns of wire. Consequently, the field winding of the series motor has a low inductance. Its impedance is therefore much lower than that of the shunt winding, and the series motor operates on ac power with better results than a shunt motor. However, the performance obtained with ac power is naturally much poorer than that obtained when the series motor is connected to a dc power source.

The performance of a series motor operating with ac power can be greatly improved by decreasing the inductance of the armature winding. This can be done by adding a new winding, called compensating winding, to the series motor. This winding is installed in the stator slots and the armature current flows through the winding. The wire loops of the compensating winding are connected so that the direction of current flow in each loop is opposite to the direction of current flow in the corresponding armature loop lying next to it, as illustrated in Figure 3-10.

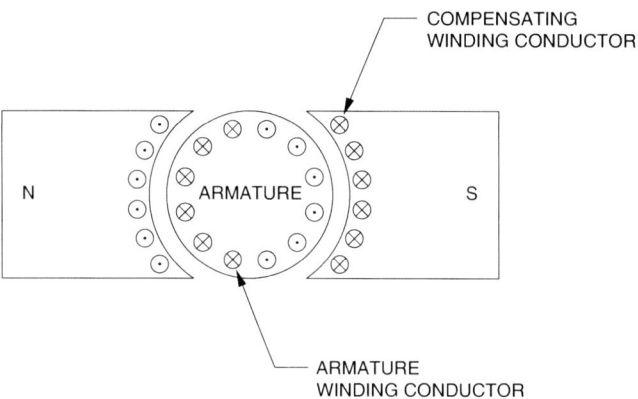

NOTE: ⊗ CURRENT INTO PAGE
⊙ CURRENT OUT OF PAGE

Figure 3-10. Current flow in the compensating winding.

This is equivalent to winding the coil of an inductor with ten turns of wire in one direction, and then ten turns of wire in the opposite direction. The resulting inductor has a very small inductance because of the cancelling effect caused by equal number of coils being wound in opposite directions. This new type of series motor is known as a universal motor because it can operate indifferently on ac power, as well as dc power.

Procedure summary

In the first part of the exercise, you will set up the equipment in the Workstation, connect the equipment as shown in Figure 3-11, and make the appropriate settings on the equipment.

In the second part of the exercise, you will change the polarities of the armature and field currents of a series motor operating on dc power and observe the effect on the direction of rotation. You will also measure the dc voltage required to make the series motor rotate at a speed of approximately 1000 r/min.

In the third part of the exercise, you will replace the dc power source with an ac power source. You will observe that the direction of rotation of the series motor can be changed by reversing the armature connections. You will measure the ac source voltage required to make the series motor rotate at a speed of approximately 1000 r/min. You will measure the armature impedance Z_A. You will compare the series motor performance obtained with dc power and ac power.

In the fourth part of the exercise, you will modify the connections to obtain the universal motor circuit shown in Figure 3-12. You will change the polarities of the armature and field currents of the universal motor operating on dc power and observe the effect on the direction of rotation. You will also measure the dc voltage required to make the universal motor rotate at a speed of approximately 1000 r/min.

In the fifth part of the exercise, you will replace the dc power source with an ac power source. You will observe that the direction of rotation of the universal motor can be changed by reversing the armature connections. You will measure the ac source voltage required to make the universal motor rotate at a speed of approximately 1000 r/min. You will measure the armature impedance Z_A. You will compare the universal motor performance obtained with dc power and ac power. You will compare the performance of the universal motor to that of the series motor.

In the sixth part of the exercise, you will add a compensating winding to the universal motor. You will observe the effect on the performance of the universal motor operating on ac power.

EQUIPMENT REQUIRED

Refer to the Equipment Utilization Chart in Appendix C to obtain the list of equipment required for this exercise.

Ex. 3-2 – The Universal Motor ♦ *Procedure*

Procedure

⚠ WARNING

High voltages are present in this laboratory exercise. Do not make or modify any banana jack connections with the power on unless otherwise specified.

Setting up the equipment

1. Install the equipment required in the EMS workstation.

   If you are performing the exercise using the EMS system, ensure that the brushes of the DC Motor/Generator are adjusted to the neutral point. To do so, connect an ac power source (terminals 4 and N of the Power Supply) to the armature of the DC Motor/Generator (terminals 1 and 2) through CURRENT INPUT I1 of the data acquisition module. Connect the shunt winding of the DC Motor/Generator (terminals 5 and 6) to VOLTAGE INPUT E1 of the Data Acquisition Interface module. Start the Metering application and open setup configuration file ACMOTOR1.DAI. Turn the Power Supply on and set the voltage control knob so that an ac current (indicated by meter I line 1) equal to half the nominal value of the armature current flows in the armature of the DC Motor/Generator. Adjust the brush adjustment lever on the DC Motor/Generator so that the voltage across the shunt winding (indicated by meter E line 1) is minimum. Turn the Power Supply off, exit the Metering application, and disconnect all leads and cable.

   Also, ensure that the brushes of the Universal Motor are adjusted to the neutral point. To do so, repeat the above procedure, connecting the series winding of the Universal Motor to VOLTAGE INPUT E1 of the data acquisition module.

   Mechanically couple the prime mover/dynamometer module to the DC Motor/Generator using a timing belt.

2. On the Power Supply, make sure the main power switch is set to the O (off) position, and the voltage control knob is turned fully counterclockwise. Ensure the Power Supply is connected to a three-phase power source.

   If you are using the Four-Quadrant Dynamometer/Power Supply, Model 8960-2, connect its POWER INPUT to a wall receptacle.

3. Ensure that the data acquisition module is connected to a USB port of the computer.

   Connect the POWER INPUT of the data acquisition module to the 24 V - AC output of the Power Supply.

   If you are using the Prime Mover/Dynamometer, Model 8960-1, connect its LOW POWER INPUT to the 24 V - AC output of the Power Supply.

Ex. 3-2 – The Universal Motor ♦ *Procedure*

On the Power Supply, set the 24 V - AC power switch to the I (on) position.

 If you are using the Four-Quadrant Dynamometer/Power Supply, Model 8960-2, turn it on by setting its POWER INPUT switch to the I (on) position. Press and hold the FUNCTION button 3 seconds to have uncorrected torque values on the display of the Four-Quadrant Dynamometer/Power Supply. The indication "NC" appears next to the function name on the display to indicate that the torque values are uncorrected.

4. Start the Data Acquisition software (LVDAC or LVDAM). Open setup configuration file DCMOTOR1.DAI.

 If you are using LVSIM-EMS in LVVL, you must use the IMPORT option in the File menu to open the configuration file.

In the Metering window, select layout 2. Make sure that the continuous refresh mode is selected.

5. Set up the series motor circuit shown in Figure 3-11.

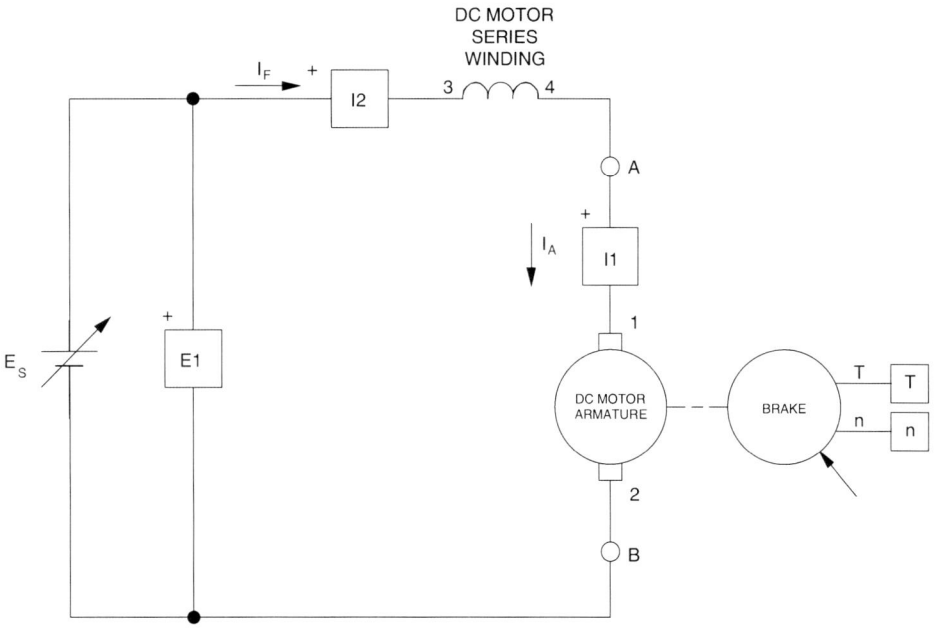

Figure 3-11. **Series motor coupled to a brake.**

6. Set the Four-Quadrant Dynamometer/Power Supply or the Prime Mover/Dynamometer to operate as a brake, then set the torque control to minimum (fully counterclockwise position). To do this, refer to Exercise 1-1 or Exercise 1-2 if necessary.

 If you are performing the exercise using LVSIM®-EMS, you can zoom in on the Prime Mover/Dynamometer before setting the controls in order to see additional front panel markings related to these controls.

Ex. 3-2 – The Universal Motor ◆ *Procedure*

Direction of rotation of a dc series motor

7. Turn the Power Supply on and slowly turn the voltage control knob until the series motor rotates at a speed of 1000 r/min ± 25 r/min. Check that both the armature current I_A and the field current I_f [indicated by meters I arm. (I_A) and I field (I_f), respectively] are positive. Record the source voltage E_S [indicated by meter E arm. (E_A)] and the direction of rotation.

   E_S = _____ V

   Direction of rotation: _____   (I_A and I_f = positive)

   Turn the Power Supply off.

8. On the Power Supply, reverse the connection of the leads at terminals 7 and N to reverse the polarity of the voltage applied to the series motor.

   Turn the Power Supply on and slightly adjust the voltage control knob until the series motor rotates at a speed of 1000 r/min ± 25 r/min. Check that both the armature current I_A and the field current I_f are negative. Record the source voltage E_S [indicated by meter E arm. (E_A)] and the direction of rotation.

   E_S = _____ V

   Direction of rotation: _____   (I_A and I_f = negative)

   Turn the Power Supply off.

   What is the direction of rotation when the armature current I_A and the field current I_f are of the same polarity?

   _____

9. Reverse the armature connection at points A and B shown in Figure 3-11.

   Turn the Power Supply on and slightly adjust the voltage control knob until the series motor rotates at a speed of 1000 r/min ± 25 r/min.

    Neglect the sign of the speed indicated by the Speed meter in the Metering window.

Check that the armature current I_A is positive and the field current I_f is negative. Record the source voltage E_S and the direction of rotation.

$E_S =$ _____ V

Direction of rotation: _____ ($I_A =$ positive, and $I_f =$ negative)

Turn the Power Supply off.

10. On the Power Supply, reverse the connection of the leads at terminals 7 and N to reverse the polarity of the voltage applied to the series motor.

    Turn the Power Supply on and slightly adjust the voltage control knob until the series motor rotates at a speed of 1000 r/min ± 25 r/min.

     Neglect the sign of the speed indicated by the Speed meter in the Metering window.

    Check that the armature current I_A is negative and the field current I_f is positive. Record the source voltage E_S and the direction of rotation.

    $E_S =$ _____ V

    Direction of rotation: _____ ($I_A =$ negative, and $I_f =$ positive)

    Turn the voltage control knob fully counterclockwise and turn the Power Supply off.

    What is the direction of rotation when the armature current I_A and the field current I_f are of opposite polarity?

    _____

    Reverse the armature connection at points A and B shown in Figure 3-11. The modules should be connected as shown in Figure 3-11.

DC series motor operating on ac power

11. Replace the variable-voltage dc source in the circuit with a variable-voltage ac source.

In the Metering window, set meters E arm. (E_A), I arm. (I_A), and I field (I_f) in the ac mode.

Turn the Power Supply on and slowly turn the voltage control knob until the series motor rotates at a speed of 1000 r/min $\pm$ 25 r/min. Record the source voltage E_S [indicated by meter E arm. (E_A)] and the direction of rotation.

E_S = _____ V

Direction of rotation: _____ (I_A and I_f of the same polarity)

Does the series motor rotate in the same direction as when it was operating on dc power with I_A and I_f of the same polarity (steps 7 and 8)?

❑ Yes ❑ No

Turn the Power Supply off.

12. Reverse the armature connection at points A and B shown in Figure 3-11.

Turn the Power Supply on and slightly adjust the voltage control knob until the series motor rotates at a speed of 1000 r/min $\pm$ 25 r/min.

> Neglect the sign of the speed indicated by the Speed meter in the Metering window.

Record the source voltage E_S and the direction of rotation.

E_S = _____ V

Direction of rotation: _____ (I_A and I_f of opposite polarity)

Does the series motor rotate in the same direction as when it was operating on dc power with I_A and I_f of the same polarity (steps 9 and 10)?

❑ Yes ❑ No

13. On the Power Supply, slowly turn the voltage control knob until the series motor stops rotating.

In the Metering window, set meter $R_A = E_A/I_A$ so that it measures impedance (Z).

Ex. 3-2 – The Universal Motor ♦ Procedure

Record in the following blank space the armature impedance Z_A of the series motor indicated by the impedance meter.

$Z_A = _____ \ \Omega$

Turn the voltage control knob fully counterclockwise and turn the Power Supply off.

Compare the dc and ac source voltages E_S required to make the series motor rotate at a speed of approximately 1000 r/min. Briefly explain why they have different values.

___

Direction of rotation of a universal motor operating on dc power

14. Remove the timing belt which couples the prime mover/dynamometer module to the DC Motor/Generator.

Mechanically couple the prime mover/dynamometer module to the Universal Motor using a timing belt.

Modify the connections to obtain the universal-motor circuit shown in Figure 3-12.

In the Metering window, set meters E arm. (E_A), I arm. (I_A), and I field (I_f) in the dc mode.

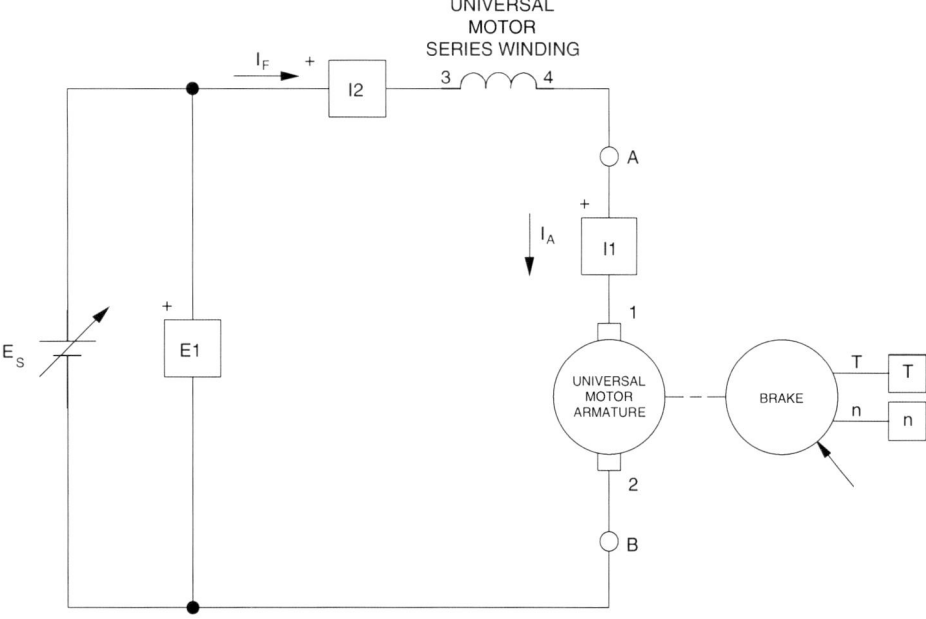

Figure 3-12. DC-powered universal motor coupled to a brake.

Ex. 3-2 – The Universal Motor ◆ *Procedure*

15. Turn the Power Supply on and slowly turn the voltage control knob until the universal motor rotates at a speed of 1000 r/min ± 25 r/min. Check that both the armature current I_A and the field current I_f are positive. Record the source voltage E_S and the direction of rotation.

$E_S = $ _____ V

Direction of rotation: _____   (I_A and I_f = positive)

Turn the Power Supply off.

16. On the Power Supply, reverse the connection of the leads at terminals 7 and N to reverse the polarity of the voltage applied to the universal motor.

Turn the Power Supply on and slightly adjust the voltage control knob until the universal motor rotates at a speed of 1000 r/min ± 25 r/min. Check that both the armature current I_A and the field current I_f are negative. Record the source voltage E_S and the direction of rotation.

$E_S = $ _____ V

Direction of rotation: _____   (I_A and I_f = negative)

Turn the Power Supply off.

What is the direction of rotation when the armature current I_A and the field current I_f are of the same polarity?

_____

17. Reverse the armature connection at points A and B shown in Figure 3-12.

Turn the Power Supply on and slightly adjust the voltage control knob until the universal motor rotates at a speed of 1000 r/min ± 25 r/min.

Neglect the sign of the speed indicated by the Speed meter in the Metering window.

Check that the armature current I_A is positive and the field current I_f is negative. Record the source voltage E_S and the direction of rotation.

$E_S = $ _____ V

Direction of rotation: _____   (I_A = positive, I_f = negative)

Turn the Power Supply off.

Ex. 3-2 – The Universal Motor ♦ *Procedure*

18. On the Power Supply, reverse the connection of the leads at terminals 7 and N to reverse the polarity of the voltage applied to the universal motor.

    Turn the Power Supply on and slightly adjust the voltage control knob until the universal motor rotates at a speed of 1000 r/min ± 25 r/min.

     Neglect the sign of the speed indicated by the Speed meter in the Metering window.

    Check that the armature current I_A is negative and the field current I_f is positive. Record the source voltage E_S and the direction of rotation.

    $E_S = $ _____ V

    Direction of rotation: _____ (I_A = negative, I_f = positive)

    Turn the voltage control knob fully counterclockwise and turn the Power Supply off.

    What is the direction of rotation when the armature current I_A and the field current I_f are of opposite polarity?

    _____

    Does a universal motor act similarly as a series motor when it is powered by a dc source?

    ❏ Yes   ❏ No

    Reverse the armature connection at points A and B shown in Figure 3-12. The modules should be connected as shown in Figure 3-12.

Universal motor operating on ac power

19. Replace the variable-voltage dc source in the circuit with a variable-voltage ac source.

    In the Metering window, set meters E arm. (E_A), I arm. (I_A), and I field (I_f) in the ac mode.

    Turn the Power Supply on and slowly turn the voltage control knob until the universal motor rotates at a speed of 1000 r/min ± 25 r/min. Record the source voltage E_S and the direction of rotation.

    $E_S = $ _____ V (without compensating winding)

    Direction of rotation: _____ (I_A and I_f of the same polarity)

Ex. 3-2 – The Universal Motor ♦ Procedure

Does the universal motor rotate in the same direction as when it was operating on dc power with I_A and I_f of the same polarity (steps 15 and 16)?

❏ Yes ❏ No

Turn the Power Supply off.

20. Reverse the armature connection at points A and B shown in Figure 3-12.

    Turn the Power Supply on and slightly adjust the voltage control knob until the universal motor rotates at a speed of 1000 r/min ± 25 r/min.

    Neglect the sign of the speed indicated by the Speed meter in the Metering window.

    Record the source voltage E_S and the direction of rotation.

    $E_S = $ _____ V (without compensating winding)

    Direction of rotation: _____ (I_A and I_f of opposite polarity)

    Does the universal motor rotate in the same direction as when it was operating on dc power with I_A and I_f of the same polarity (steps 17 and 18)?

    ❏ Yes ❏ No

21. On the Power Supply, slowly turn the voltage control knob until the universal motor stops rotating.

    Record in the following blank space the armature impedance Z_A of the universal motor indicated by the impedance meter.

    $Z_A = $ _____ Ω (without compensating winding)

    Turn the voltage control knob fully counterclockwise and turn the Power Supply off.

    Compare the dc and ac source voltages E_S required to make the universal motor rotate at a speed of approximately 1000 r/min. Briefly explain why they have different values.

    _____
    _____
    _____

Compare the source voltages required to make the series motor and the universal motor (without compensating winding) rotate at a speed of approximately 1000 r/min when they operate on dc power and on ac power. Do both motors require a higher source voltage when they operate on ac power?

❏ Yes   ❏ No

Reverse the armature connection at points A and B shown in Figure 3-12.

Effect of the compensating winding

22. Modify the connections to connect the compensating winding of the Universal Motor as shown in Figure 3-13.

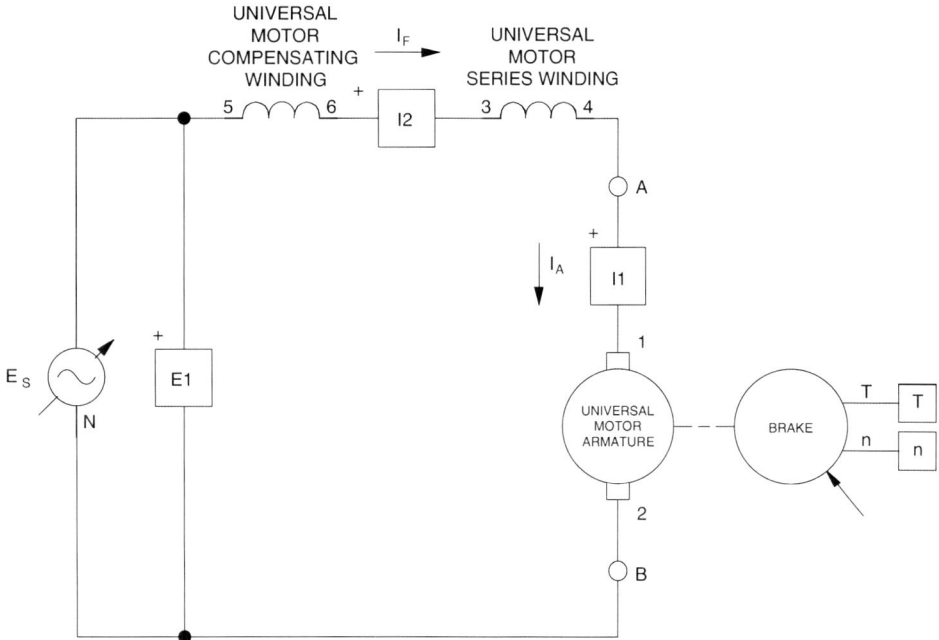

Figure 3-13. AC-powered universal motor (with compensating winding) coupled to a brake.

Turn the Power Supply on and slowly turn the voltage control knob until the universal motor rotates at a speed of 1000 r/min ± 25 r/min. Record the source voltage E_S.

$E_S = $ _____ V (with compensating winding)

On the Power Supply, slowly turn the voltage control knob until the universal motor stops rotating.

Record in the following blank space the armature impedance Z_A of the universal motor indicated by the impedance meter.

$Z_A = $ _____ Ω (with compensating winding)

Turn the voltage control knob fully counterclockwise and turn the Power Supply off.

Compare the ac source voltages E_S required to make the universal motors with and without compensating winding rotate at a speed of approximately 1000 r/min. Briefly explain why they have different values.

___

___

___

23. On the Power Supply, set the 24 V - AC power switch to the O (off) position

If you are using the Four-Quadrant Dynamometer/Power Supply, Model 8960-2, turn it off by setting its POWER INPUT switch to the O (off) position.

Remove all leads and cables.

Additional experiments

Speed versus torque characteristic of an ac-powered universal motor

You can obtain the speed versus torque characteristic of a universal motor (with a compensating winding) powered by an ac source. To do so, make sure the Power Supply is turned off and set up the universal motor circuit shown in Figure 3-13. Set the brake torque control to minimum (fully counterclockwise position). (If you are using the Four-Quadrant Dynamometer/Power Supply, make sure to press and hold the FUNCTION button 3 seconds to have uncorrected torque values on this module's display). In the Metering window, make sure that meters E arm. (E_A), I arm. (I_A), and I field (I_f) are in the ac mode and the torque correction function of the Torque meter is enabled. Clear the data (if any) recorded in the Data Table. Turn the Power Supply on and set the ac source voltage E_S to the nominal voltage of the Universal Motor. On the dynamometer, set the torque control so that the torque indicated by the Torque meter in the Metering window increases by 0.2 N·m (2.0 lbf·in) increments up to about 2.3 N·m (about 20.0 lbf·in). For each torque setting, readjust the voltage control knob of the Power Supply so that the armature voltage E_A remains equal to the value set previously, wait until the motor speed stabilizes, and then record the data in the Data Table. When all data has been recorded, turn the Power Supply off. Edit the Data Table so as to keep only the values of the speed n, torque T, armature current I_A, and source voltage E_S. Entitle the Data Table as DT321. Plot a graph of the speed (obtained from the Speed meter) as a function of the torque (obtained from the Torque meter). Entitle the graph as G321. Compare the speed versus torque characteristic of the universal motor (graph G321) to that of the dc series motor (graph G223 obtained in Exercise 2-2).

The armature current may exceed the rated value while performing this manipulation. It is, therefore, suggested to complete the manipulation within a time interval of 5 minutes or less.

Ex. 3-2 – The Universal Motor ♦ *Conclusion*

DC shunt motor operating on ac power

You can observe the operation of a shunt motor connected to an ac power source. To do so, make sure the Power Supply is turned off and set up the shunt motor circuit shown in Figure 3-14. In the Metering window, make sure that meters E arm. (E_A), I arm. (I_A), and I field (I_f) are in the ac mode. Turn the Power Supply on and turn the voltage control knob until the shunt motor starts to rotate. Note the direction of rotation. Turn the Power Supply off. Reverse the armature connection at points A and B shown in Figure 3-14. Turn the Power Supply on and turn the voltage control knob until the shunt motor starts to rotate. Note the direction of rotation. Turn the Power Supply off.

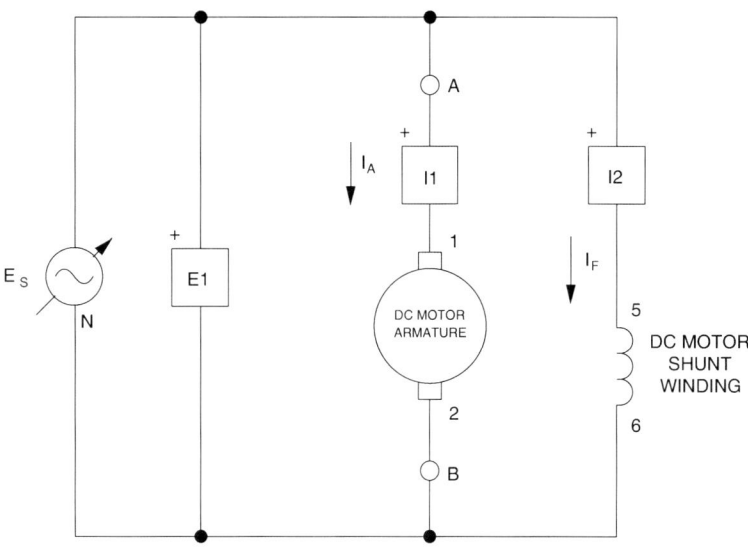

Figure 3-14. AC-powered shunt motor.

CONCLUSION

In this exercise, you demonstrated that the dc series motor and the universal motor without compensating winding have similar operation whether they are supplied with dc or ac power. You observed that the direction of rotation of these motors depends on the polarities of the armature and field currents. You found that the performance of these motors is rather poor when they operate on ac power because their armature impedance Z_A is fairly high. You observed that the performance of a universal motor operating on ac power can be greatly improved by adding a compensating winding that reduces the armature impedance Z_A.

If you have performed the additional experiments, you plotted a graph of speed versus torque for an ac-powered universal motor with a compensating winding. You found that the speed-torque characteristic of this motor is similar to that of the dc series motor, i.e., the speed decreases rapidly and non-linearly as the torque increases. You verified that a dc shunt motor can operate on ac power.

Ex. 3-2 – The Universal Motor ♦ *Review Questions*

REVIEW QUESTIONS

1. What effect does reversing the dc power connections to a series motor have on its direction of rotation?

    a. It remains clockwise.
    b. It reverses.
    c. It changes from clockwise to counterclockwise.
    d. It remains the same.

2. What effect does reversing the armature winding connections to a series motor have on its direction of rotation?

    a. It remains counterclockwise.
    b. It reverses.
    c. It changes from counterclockwise to clockwise.
    d. It remains the same.

3. A universal motor is a dc series motor

    a. that operates on ac power only.
    b. with a compensating winding that operates on dc power only.
    c. with a compensating winding that operates on ac power only.
    d. with a compensating winding that operates on either ac and dc power.

4. The compensating winding in a universal motor helps reducing

    a. the motor torque T.
    b. the armature current I_A.
    c. the armature impedance Z_A.
    d. the motor speed n.

5. If the armature connections to a universal motor are reversed, the

    a. armature current will decrease.
    b. direction of rotation will reverse.
    c. motor will stop rotating.
    d. motor speed will become unstable.

Unit Test

1. Armature reaction in a dc machine

   a. is due to an increase of the armature voltage.
   b. occurs when the motor is connected to an ac power source.
   c. occurs when the motor is connected to a dc power source.
   d. is due to an increase of the armature current.

2. Armature reaction modifies the characteristics of a dc machine because

   a. it increases wear on the brushes and the commutator.
   b. it affects the magnetic field produced by the field electromagnet.
   c. it causes saturation.
   d. Both a and b.

3. Armature reaction causes the output voltage of a dc generator to decrease because

   a. it increases wear on the brushes and the commutator.
   b. it causes saturation.
   c. it delays the voltage induced across the armature winding.
   d. All of the above.

4. A permanent-magnet dc motor has better characteristics than the separately-excited, shunt, and series motors because

   a. the magnetic field produced by the permanent magnet is so strong that it cannot be affected significantly by another magnetic source.
   b. it has a low armature inductance.
   c. it has a compensating winding.
   d. Both a and b.

5. Saturation occurs in a dc machine when

   a. the armature voltage increases and exceeds the nominal value.
   b. the motor is connected to an ac power source.
   c. the field current exceeds the nominal value.
   d. the armature current exceeds the nominal value.

6. The nominal value of the field current of a dc machine is chosen to be at the beginning of the saturation knee

   a. to ensure that the speed versus voltage characteristic is linear.
   b. to ensure that the torque versus current characteristic is linear.
   c. to obtain as much torque as possible with a field current that is as low as possible.
   d. Both a and b.

7. Why is it nearly impossible to obtain satisfactory performance from a shunt motor connected to an ac power source?

   a. Because the shunt winding consist of a large number of turns.
   b. Because the shunt winding has a large inductance.
   c. Because it is difficult for an alternating current to flow in the shunt winding.
   d. All of the above.

8. The direction of rotation of a dc series motor or a universal motor connected to a dc power source

   a. depends on the polarities of the armature and field currents.
   b. depends exclusively on the polarity of the armature current.
   c. depends exclusively on the polarity of the field current.
   d. depends on the connection of the compensating winding.

9. The ac voltage required to make a series motor rotate at a given speed is higher than the dc voltage required to make the same motor rotate at the same speed. This is because

   a. armature reaction occurs when the motor operate on ac power.
   b. the armature impedance of the motor is fairly high.
   c. saturation occurs when the motor operate on ac power.
   d. Both a and b.

10. The performance of a series motor operating on ac power can be improved by

    a. adding a compensating winding that increases the armature reactance.
    b. adding permanent magnets.
    c. adding a compensating winding that decreases the armature reactance.
    d. None of the above

Unit 4

AC Induction Motors

UNIT OBJECTIVE

When you have completed this unit, you will be able to demonstrate and explain the operation of ac induction motors using the Squirrel-Cage Induction Motor module and the Capacitor-Start Motor module.

DISCUSSION OF FUNDAMENTALS

As you saw in Unit 1, a voltage is induced between the ends of a wire loop when the magnetic flux linking the loop varies as a function of time. If the ends of the wire loop are short-circuited together, a current flows in the loop. Figure 4-1 shows a magnet that is displaced rapidly towards the right above a group of conductors. The conductors are short-circuited at their extremities by bars A and B and form a type of ladder.

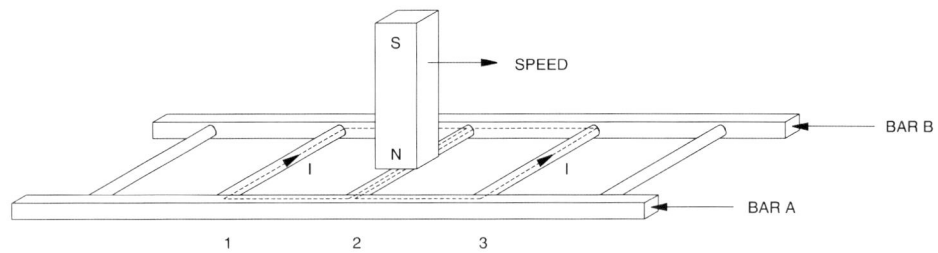

Figure 4-1. Magnet moving above a conducting ladder.

Current flows in the loop formed by conductors 1 and 2, as well as in the loop formed by conductors 2 and 3. These currents create magnetic fields with north and south poles as shown in Figure 4-2.

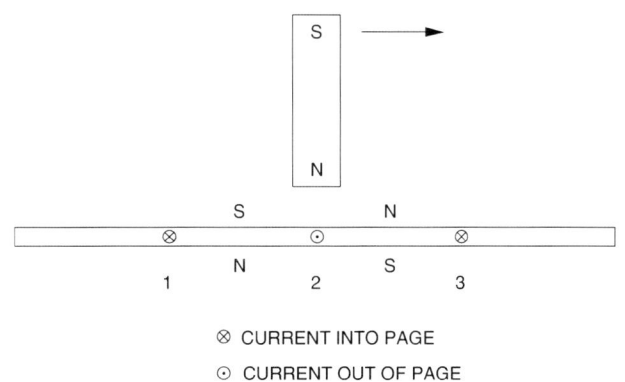

Figure 4-2. Current in the conductors creates magnetic fields.

The interaction between the magnetic field of the magnet and the magnetic fields produced by the currents induced in the ladder creates a force between the moving magnet and the electromagnet (the conducting ladder). This force causes the ladder to be pulled along in the direction of the moving magnet. However, if the ladder moves at the same speed as the magnet, there is no longer a variation in the magnetic flux. Consequently, there is no induced voltage to cause current flow in the wire loops, meaning that there is no longer a magnetic force acting on the ladder. Therefore, the ladder must move at a speed which is lower than that of the moving magnet for a magnetic force to pull the ladder in the direction of the moving magnet. The greater the speed difference between the two, the greater the variation in magnetic flux, and therefore, the greater the magnetic force acting on the conducting ladder.

The rotor of an asynchronous induction motor is made by closing a ladder similar to that shown in Figure 4-1 upon itself to form a type of squirrel cage as shown in Figure 4-3. This is where the name squirrel-cage induction motor comes from.

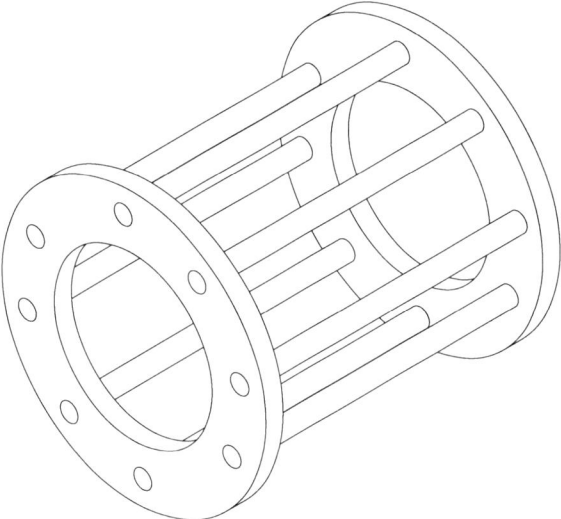

Figure 4-3. Closing a ladder upon itself forms a squirrel cage.

To make it easier for the magnetic flux to circulate, the rotor of a squirrel-cage induction motor is placed inside a laminated iron cylinder. The stator of the induction motor acts as a rotating electromagnet. The rotating electromagnet causes torque which pulls the rotor along in much the same manner as the moving magnet in Figure 4-1 pulls the ladder.

Exercise 4-1

The Three-Phase Squirrel-Cage Induction Motor

EXERCISE OBJECTIVE

When you have completed this exercise, you will be able to demonstrate the operating characteristics of a three-phase induction motor using the Four-Pole Squirrel-Cage Induction Motor module.

DISCUSSION

One of the ways of creating a rotating electromagnet is to connect a three-phase power source to a stator made of three electromagnets A, B, and C, that are placed at 120° to one another as shown in Figure 4-4.

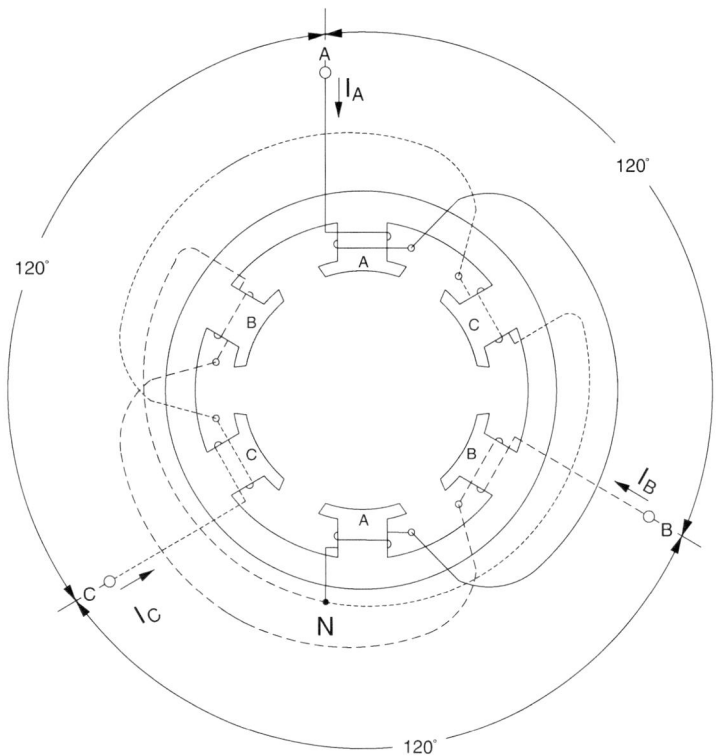

Figure 4-4. Three-phase stator windings.

Ex. 4-1 – The Three-Phase Squirrel-Cage Induction Motor ♦ *Discussion*

When sine-wave currents that are phase shifted of 120° to each other, like those shown in Figure 4-5, are flowing in stator electromagnets A, B, and C, a magnetic field that rotates very regularly is obtained.

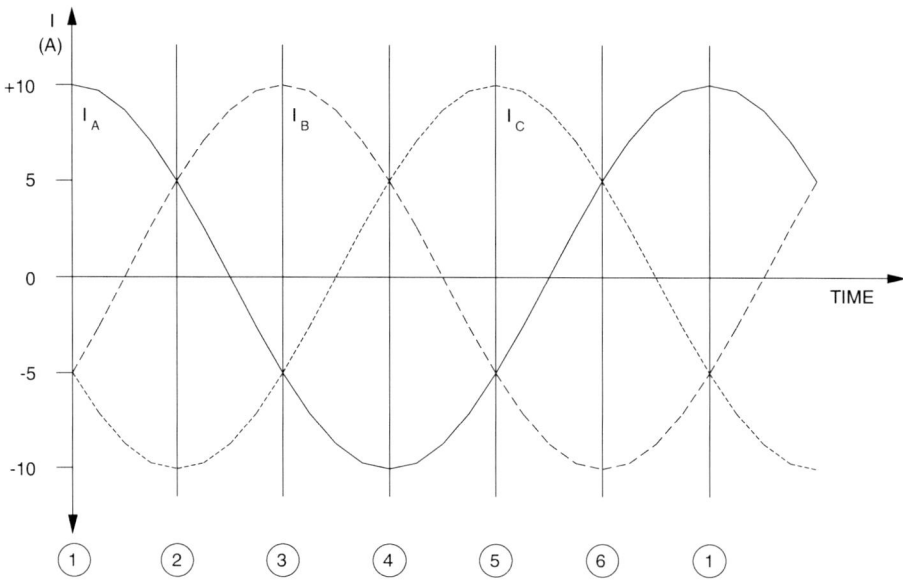

Figure 4-5. Three-phase sine-wave currents flowing in the stator windings.

Figure 4-6 illustrates the magnetic field created by stator electromagnets A, B, and C at instants numbered 1 to 6 in Figure 4-5. Notice that the magnetic lines of force exit at the north pole of each electromagnet and enter at the south pole. As can be seen, the magnetic field rotates clockwise.

The use of sine-wave currents produces a magnetic field that rotates regularly and whose strength does not vary over time. The speed of the rotating magnetic field is known as the synchronous speed (n_S) and is proportional to the frequency of the ac power source. A rotating magnetic field can also be obtained using other combinations of sine-wave currents that are phase-shifted with respect to each other, but three-phase sine-wave currents are used more frequently.

When a squirrel-cage rotor is placed inside a rotating magnetic field, it is pulled around in the same direction as the rotating field. Interchanging the power connections to two of the stator windings (interchanging A with B for example) interchanges two of the three currents and reverses the phase sequence. This causes the rotating field to reverse direction. As a result, the direction of rotation of the motor is also reversed.

Ex. 4-1 – The Three-Phase Squirrel-Cage Induction Motor ♦ *Discussion*

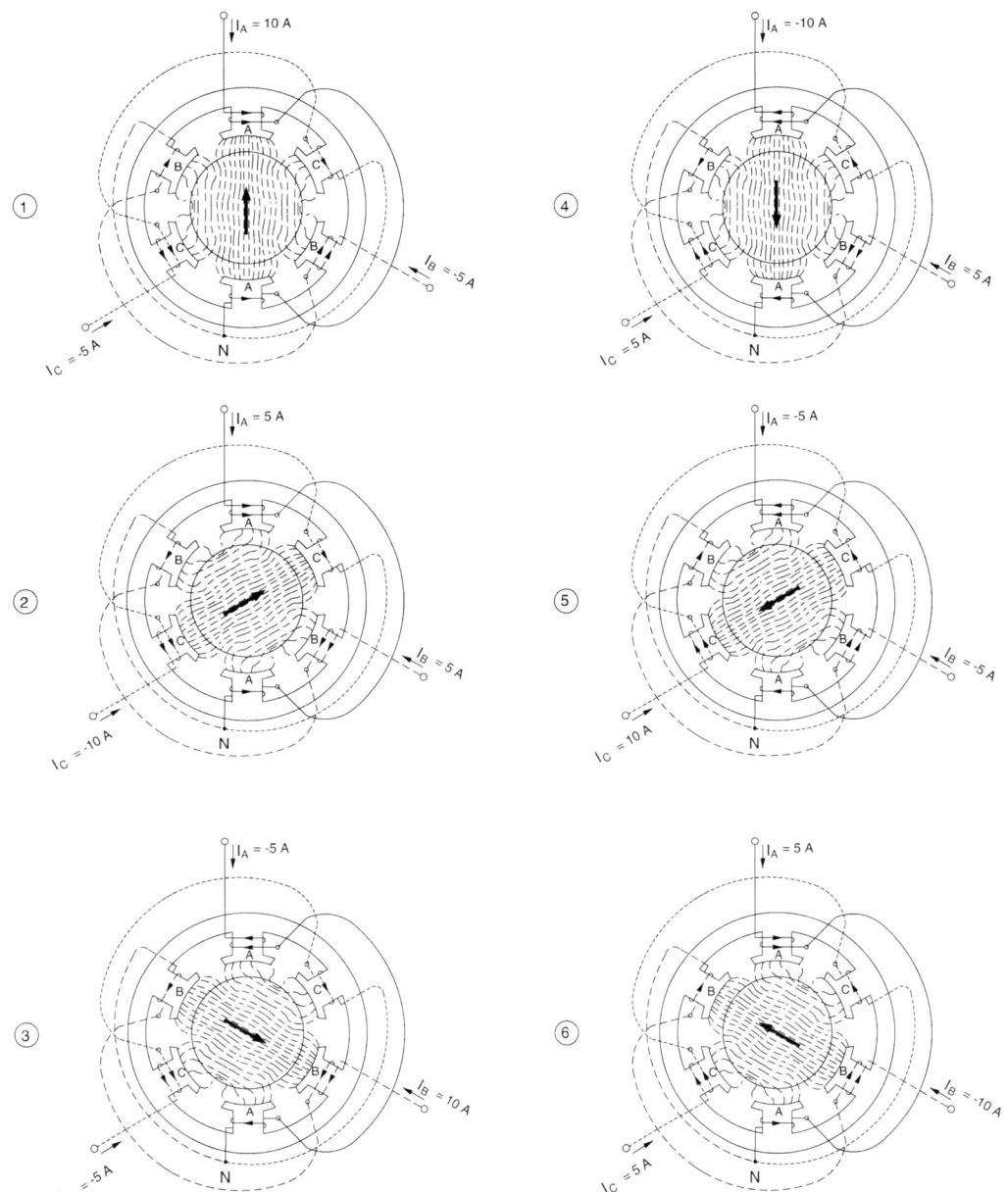

Figure 4-6. Position of the rotating magnetic field at various instants. (From Electrical Machines, Drives, and Power Systems by Theodore Wildi. Copyright © 1991, 1981 Sperika Enterprises Ltd. © Published by Prentice Hall. All rights reserved.).

Referring to what has been said in the Discussion of Fundamentals of this unit, one can easily deduce that the torque produced by a squirrel-cage induction motor increases as the difference in speed between the rotating magnetic field and the rotor increases. The difference in speed between the two is called slip. A plot of the speed versus torque characteristic for a squirrel-cage induction motor gives a curve similar to that shown in Figure 4-7. As can be seen, the motor speed (rotor speed) is always lower than the synchronous speed n_S because slip is necessary for the motor to develop torque. The synchronous speed for the motors is 1800 r/min for 60 Hz power, and 1500 r/min for 50-Hz power.

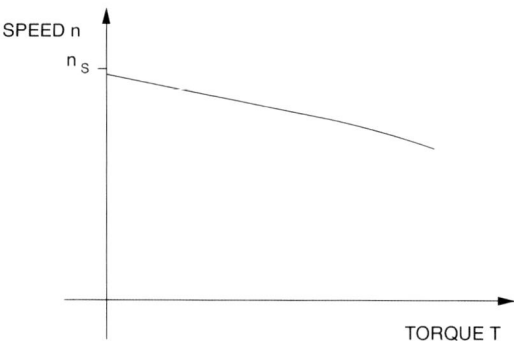

Figure 4-7. Speed versus torque characteristic of a squirrel-cage induction motor.

The speed versus torque characteristic of the squirrel-cage induction motor is very similar to that obtained previously for a separately-excited dc motor. However, the currents induced in the squirrel-cage rotor must change direction more and more rapidly as the slip increases. In other words, the frequency of the currents induced in the rotor increases as the slip increases. Since the rotor is made up of iron and coils of wire, it has an inductance that opposes rapid changes in current. As a result, the currents induced in the rotor are no longer directly proportional to the slip of the motor. This affects the speed versus torque characteristic as shown in Figure 4-8.

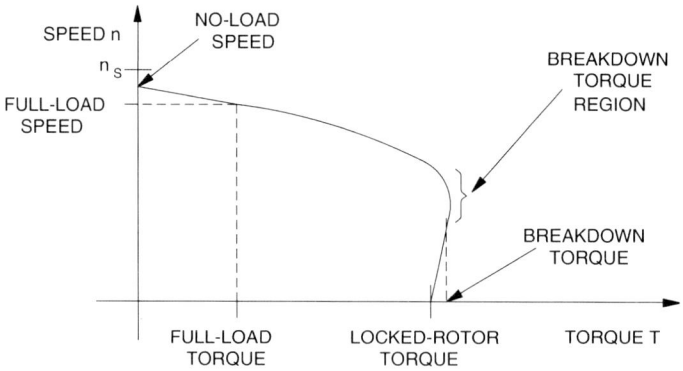

Figure 4-8. The motor inductance affects the speed versus torque characteristic.

As the curve shows, the no-load speed is slightly less than the synchronous speed n_S, but as the load torque increases, motor speed decreases. For the nominal value of motor torque (full-load torque) corresponds a nominal operating speed (full-load speed). Further increases in load torque lead to a point of instability, called breakdown torque, after which both motor speed and output torque decrease. The torque value at zero speed, called locked-rotor torque, is often less than the breakdown torque. At start-up, and at low speed, motor current is very high and the amount of power that is consumed is higher than during normal operation.

Another characteristic of three-phase squirrel-cage induction motors is the fact that they always draw reactive power from the ac power source. The reactive power even exceeds the active power when the squirrel-cage induction motor rotates without load. The reactive power is necessary to create the magnetic field in the machine in the same way that an inductor needs reactive power to create the magnetic field surrounding the inductor.

Procedure summary

In the first part of the exercise, you will set up the equipment in the Workstation, connect the equipment as shown in Figure 4-9, and make the appropriate settings on the equipment.

In the second part of the exercise, you will apply the nominal line voltage to the squirrel-cage induction motor, note the motor direction of rotation, and measure the motor no-load speed. You will then increase the mechanical load applied to the squirrel-cage induction motor by steps. For each step, you will record in the Data Table various electrical and mechanical parameters related to the motor. You will then use this data to plot various graphs and determine many of the characteristics of the squirrel-cage induction motor.

In the third part of the exercise, you will interchange two of the leads that supply power to the squirrel-cage induction motor and observe if this affects the direction of rotation.

EQUIPMENT REQUIRED

Refer to the Equipment Utilization Chart in Appendix C to obtain the list of equipment required for this exercise.

PROCEDURE

 High voltages are present in this laboratory exercise. Do not make or modify any banana jack connections with the power on unless otherwise specified.

Setting up the equipment

1. Install the equipment required in the EMS workstation.

   Mechanically couple the prime mover/dynamometer module to the Four-Pole Squirrel-Cage Induction Motor.

2. On the Power Supply, make sure the main power switch is set to the O (off) position, and the voltage control knob is turned fully counterclockwise. Ensure the Power Supply is connected to a three-phase power source.

    If you are using the Four-Quadrant Dynamometer/Power Supply, Model 8960-2, connect its POWER INPUT to a wall receptacle.

3. Ensure that the data acquisition module is connected to a USB port of the computer.

   Connect the POWER INPUT of the data acquisition module to the 24 V - AC output of the Power Supply.

    If you are using the Prime Mover/Dynamometer, Model 8960-1, connect its LOW POWER INPUT to the 24 V - AC output of the Power Supply.

   On the Power Supply, set the 24 V - AC power switch to the I (on) position.

    If you are using the Four-Quadrant Dynamometer/Power Supply, Model 8960-2, turn it on by setting its POWER INPUT switch to the I (on) position. Press and hold the FUNCTION button 3 seconds to have uncorrected torque values on the display of the Four-Quadrant Dynamometer/Power Supply. The indication "NC" appears next to the function name on the display to indicate that the torque values are uncorrected.

4. Start the Data Acquisition software (LVDAC or LVDAM). Open setup configuration file ACMOTOR1.DAI.

    If you are using LVSIM-EMS in LVVL, you must use the IMPORT option in the File menu to open the configuration file.

   In the Metering window, select layout 2. Make sure that the continuous refresh mode is selected.

5. Connect the equipment as shown in Figure 4-9.

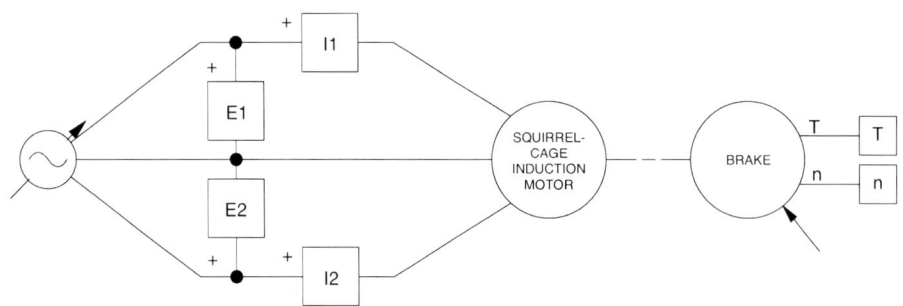

Figure 4-9. Squirrel-cage induction motor coupled to a brake.

6. Set the Four-Quadrant Dynamometer/Power Supply or the Prime Mover/Dynamometer to operate as a brake, then set the brake torque control to minimum (fully counterclockwise position). To do this, refer to Exercise 1-1 or Exercise 1-2 if necessary.

    If you are performing the exercise using LVSIM®-EMS, you can zoom in on the Prime Mover/Dynamometer before setting the controls in order to see additional front panels markings related to these controls.

Characteristics of a squirrel-cage induction motor

7. Turn the Power Supply on and set the voltage control knob so that the line voltage indicated by meter E line 1 is equal to the nominal line voltage of the squirrel-cage induction motor.

    The rating of any of the machines is indicated in the lower left corner of the module front panel.

   What is the direction of rotation of the squirrel-cage induction motor?

   ___

   Record in the following blank space the motor speed indicated by the Speed meter in the Metering window.

   $n =$ _____ r/min

   Is the no-load speed almost equal to the speed of the rotating magnetic field (synchronous speed) given in the Discussion?

   ❑ Yes   ❑ No

8. In the Metering window, make sure that the torque correction function of the Torque meter is enabled. The Torque meter indicates the output torque of the squirrel-cage induction motor.

   On the brake, adjust the torque control so that the mechanical power developed by the squirrel-cage induction motor (indicated by meter Mech. Power in the Metering window) is equal to 175 W (nominal motor output power).

   Record the nominal speed $n_{NOM.}$, torque $T_{NOM.}$, and line current $I_{NOM.}$ of the squirrel-cage induction motor in the following blank spaces. These parameters are indicated by meters Speed, Torque, and I line 1, respectively.

   $n_{NOM.} =$ _____ r/min

   $T_{NOM.} =$ _____ N·m (lbf·in)

   $I_{NOM.} =$ _____ A

   On the brake, set the torque control to minimum (fully counterclockwise position). The torque indicated by the brake's display should be 0 N·m (0 lbf·in).

9. Record the motor line voltage E_{LINE}, line current I_{LINE}, active power P, reactive power Q, output torque T, and speed n in the Data Table. These parameters are indicated by meters E line 1, I line 1, Act. power, React. power, Speed, and Torque, respectively.

On the brake, carefully adjust the torque control so that the torque indicated by the Torque meter increases by 0.3 N·m (3.0 lbf·in) increments up to about 2.1 N·m (19 lbf·in). For each torque setting, record the data in the Data Table.

On the brake, continue to adjust the torque control so that the torque indicated by the Torque meter increases by 0.1 N·m (1.0 lbf·in) increments, until the motor speed starts to decrease fairly rapidly (breakdown torque region). For each torque setting, record the data in the Data Table.

Once the motor speed has stabilized, record the data in the Data Table.

The nominal line current of the Four-Pole Squirrel-Cage Induction Motor may be exceeded while performing this manipulation. It is, therefore, suggested to complete the manipulation within a time interval of 5 minutes or less.

10. When all data has been recorded, set the torque control knob on the brake to minimum (fully counterclockwise), turn the voltage control knob fully counterclockwise, and turn the Power Supply off.

In the Data Table window, confirm that the data has been stored, entitle the Data Table as DT411, and print the Data Table.

Refer to the user guide dealing with the computer-based instruments for EMS to know how to edit, entitle, and print a Data Table.

Does the motor line current indicated in column I line 1 increase as the mechanical load applied to the squirrel-cage induction motor increases?

❏ Yes   ❏ No

11. In the Graph window, make the appropriate settings to obtain a graph of the motor speed (obtained from the Speed meter) as a function of the motor torque (obtained from the Torque meter). Entitle the graph as G411, name the x-axis as Squirrel-cage induction-motor torque, name the y-axis as Squirrel-cage induction-motor speed, and print the graph.

Refer to the user guide dealing with the computer-based instruments for EMS to know how to use the Graph window of the Metering application to obtain a graph, entitle a graph, name the axes of a graph, and print a graph.

Briefly describe how the speed varies as the mechanical load applied to the squirrel-cage induction motor increases, i.e., as the motor torque increases.

_____

_____

Ex. 4-1 – The Three-Phase Squirrel-Cage Induction Motor ◆ *Procedure*

12. Indicate on graph G411 the nominal speed and torque of the squirrel-cage induction motor measured previously.

    Determine the breakdown torque of the squirrel-cage induction motor using graph G411.

    $T_{BREAKDOWN}$ = _____ N·m (lbf·in)

    Determine the minimum-speed torque using graph G411. This torque is a good approximation of the locked-rotor torque of the squirrel-cage induction motor.

    $T_{LOCKED\ ROTOR}$ = _____ N·m (lbf·in)

    Compare the breakdown torque and locked-rotor torque with the nominal torque of the squirrel-cage induction motor.

    _____

    _____

    _____

13. In the Graph window, make the appropriate settings to obtain a graph of the motor active (P) and reactive (Q) power (obtained from meters Act. power and React. power, respectively) as a function of the motor speed (obtained from the Speed meter) using the data recorded previously in the Data Table (DT411). Entitle the graph as G411-1, name the x-axis as Squirrel-cage induction-motor speed, name the y-axis as Squirrel-cage induction-motor active power and reactive power, and print the graph.

    Does graph G411-1 confirm that the squirrel-cage induction motor always draws reactive power from the ac power source?

    ❑ Yes   ❑ No

    Does graph G411-1 confirm that the squirrel-cage induction motor draws more electrical power from the ac power source as it drives an heavier load?

    ❑ Yes   ❑ No

    Observe that when the squirrel-cage induction motor rotates without load, the reactive power exceeds the active power. What does this reveal?

    _____

    _____

Ex. 4-1 – The Three-Phase Squirrel-Cage Induction Motor ◆ *Procedure*

14. In the Graph window, make the appropriate settings to obtain a graph of the motor line current I_{LINE} (obtained from meter I line 1) as a function of the motor speed (obtained from the Speed meter) using the data recorded previously in the Data Table (DT411). Entitle the graph as G411-2, name the x-axis as Squirrel-cage induction-motor speed, name the y-axis as Squirrel-cage induction-motor line current, and print the graph.

    How does the line current vary as the motor speed decreases?

    _____

15. Indicate on graph G411-2 the nominal line current of the squirrel-cage induction motor measured previously.

    By how many times is the starting line current greater than the nominal line current? (Use the line current measured at minimum speed as the starting current.)

    _____

    _____

Direction of rotation

16. On the Four-Pole Squirrel-Cage Induction Motor, interchange any two of the three leads connected to the stator windings.

    Turn the Power Supply on and set the voltage control knob so that the line voltage indicated by meter E line 1 is approximately equal to the nominal line voltage of the squirrel-cage induction motor.

    What is the direction of rotation of the squirrel-cage induction motor?

    _____

    Does the squirrel-cage induction motor rotate opposite to the direction noted previously in this exercise?

    ❑ Yes   ❑ No

17. On the Power Supply, set the 24 V - AC power switch to the O (off) position.

    If you are using the Four-Quadrant Dynamometer/Power Supply, Model 8960-2, turn it off by setting its POWER INPUT switch to the O (off) position.

    Remove all leads and cables.

Ex. 4-1 – The Three-Phase Squirrel-Cage Induction Motor ♦ *Conclusion*

CONCLUSION

In this exercise, you observed that when the nominal line voltage is applied to the stator windings of a squirrel-cage induction motor without mechanical load, the rotor turns at approximately the same speed as the rotating magnetic field (synchronous speed). You saw that interchanging any two of the three leads supplying power to the stator windings reverses the phase sequence, and thereby, causes the motor to rotate in the opposite direction. You observed that the motor line currents increase as the mechanical load increases, thus showing that the squirrel-cage induction motor requires more electric power to drive heavier loads. You plotted a graph of speed versus torque and used it to determine the nominal, breakdown, and locked-rotor torques of the squirrel-cage induction motor. You also plotted a graph of the motor reactive power versus speed and observed that the squirrel-cage induction motor draws reactive power from the ac power source to create its magnetic field. Finally, you plotted a graph of the motor line current versus speed and observed that the starting current is many times greater than the nominal line current.

REVIEW QUESTIONS

1. The speed of the rotating magnetic field created by three-phase power is called

   a. no-load speed.
   b. synchronous speed.
   c. slip speed.
   d. nominal speed.

2. The difference between the synchronous speed and the rotation speed of a squirrel-cage induction motor is

   a. known as slip.
   b. always greater than 10%.
   c. known as slip torque.
   d. always less than 1%.

3. Reactive power is consumed by a squirrel-cage induction motor because

   a. it uses three-phase power.
   b. it does not require active power.
   c. it requires reactive power to create the rotating magnetic field.
   d. it has a squirrel-cage.

4. Does the speed of a squirrel-cage induction motor increase or decrease when the motor load increases?

   a. It increases.
   b. It decreases.
   c. It stays the same because speed is independent of motor load.
   d. The speed oscillates around the original value.

5. What happens when two of the three leads supplying power to a squirrel-cage induction motor are reversed?

   a. The motor does not start.
   b. Nothing.
   c. The motor reverses its direction of rotation.
   d. The motor consumes more reactive power.

Exercise 4-2

Eddy-Current Brake and Asynchronous Generator

EXERCISE OBJECTIVE   When you have completed this exercise, you will be able to demonstrate the main operating characteristics of an eddy-current brake as well as those of an asynchronous generator using the Four-Pole Squirrel-Cage Induction Motor and prime mover/dynamometer module.

DISCUSSION   Figure 4-10 illustrates the magnet and the conducting ladder shown previously in Figure 4-1. This time, however, the magnet is fixed and the ladder is displaced rapidly towards right.

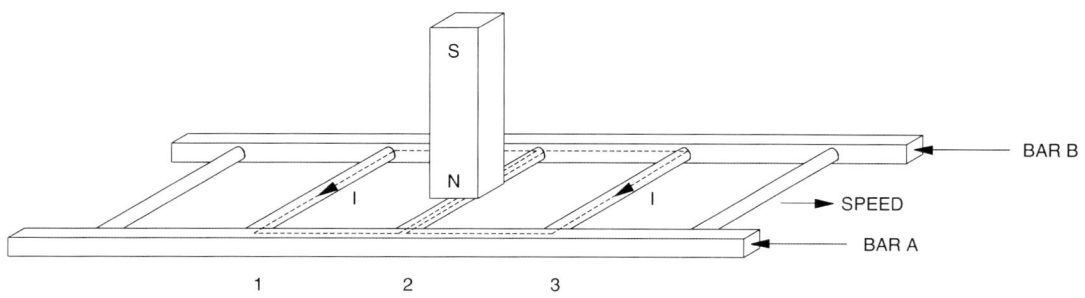

Figure 4-10. Conducting ladder moving below a magnet.

This causes current to flow in the loop formed by conductors 1 and 2, as well as in the loop formed by conductors 2 and 3. These currents create magnetic fields with north and south poles, as shown in Figure 4-11.

The interaction between the magnetic field of the magnet and the magnetic fields produced by the currents induced in the ladder creates a force between the fixed magnet and the moving electromagnet (the conducting ladder). This force causes the ladder to be pulled along in the direction of the fixed magnet, and thereby, tends to reduce the ladder speed. However, if the ladder stops moving, there is no longer a variation in the magnetic flux. Consequently, there is no induced voltage to cause current flow in the wire loops, meaning that there is no longer a magnetic force acting on the ladder. Therefore, a magnetic braking force acts on the ladder as long as it is moving. The greater the ladder speed (up to a certain limit), the greater the variation in magnetic flux, and therefore, the greater the magnetic braking force acting on the conducting ladder.

Ex. 4-2 – Eddy-Current Brake and Asynchronous Generator ♦ *Discussion*

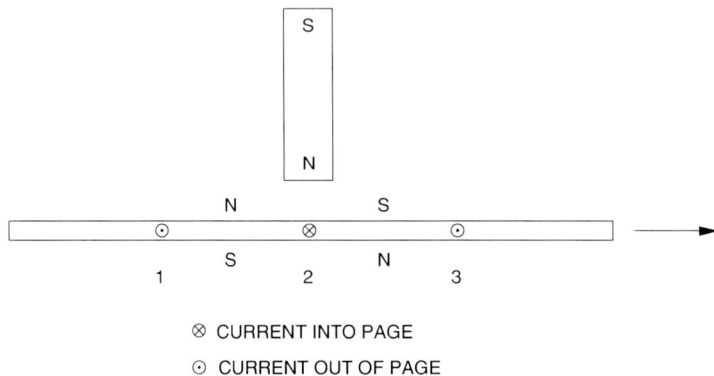

⊗ CURRENT INTO PAGE
⊙ CURRENT OUT OF PAGE

Figure 4-11. Current in the conductors creates magnetic fields.

The above principle is used to advantage in the eddy-current brake, in which a fixed (stator) electromagnet creates a braking torque that acts on a squirrel-cage rotor obtained by closing a ladder similar to that shown in Figure 4-10 upon itself. Figure 4-12 illustrates an eddy-current brake. Notice that a variable-voltage dc source is used to make current flow in the stator electromagnet. Varying the dc source voltage allows variation of the current in the electromagnet, and thereby, variation of the electromagnet strength. The greater the electromagnet strength, the greater the magnetic flux in the machine, the greater the currents induced in the squirrel-cage rotor as it turns, and the greater the braking force.

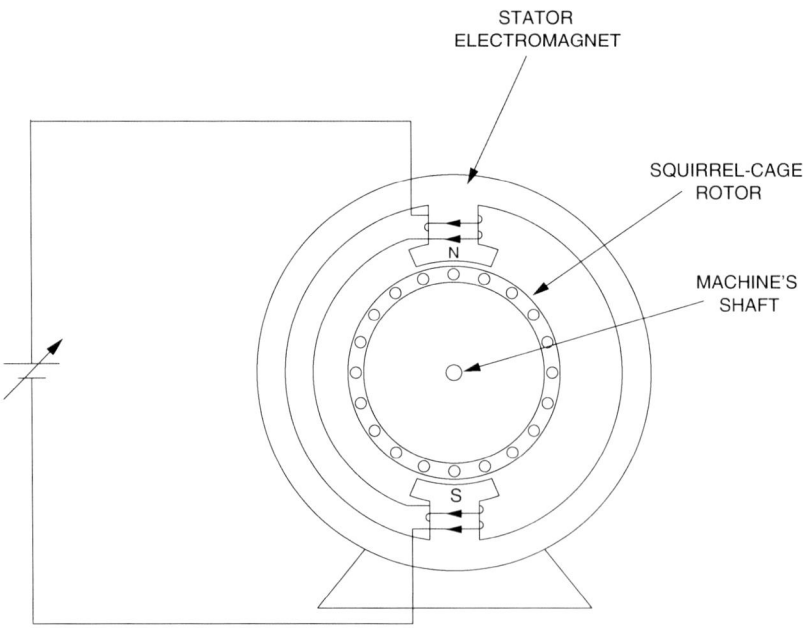

Figure 4-12. Eddy-current brake.

Note that mechanical energy from the driving machine is transferred to the eddy-current brake during braking. This energy is converted to electrical energy that is dissipated as heat in the squirrel-cage rotor of the eddy-current brake.

A braking force like that in eddy-current brake can be produced in squirrel-cage induction motors. This occurs when the rotor rotates at a speed higher than that of the rotating magnetic field (synchronous speed n_S). This is equivalent to having a fixed magnet and a moving ladder as in Figure 4-10. As in the eddy-current brake, mechanical energy is converted to electrical energy when the speed of a squirrel-cage induction motor is higher than the synchronous speed n_S. However, most of this energy is not dissipated as heat in the rotor of the squirrel-cage induction motor. It is returned to the ac power source connected to the stator windings of the motor. Therefore, a squirrel-cage induction motor operates as an asynchronous generator when its speed is higher than the synchronous speed n_S.

In brief, when the rotor of a squirrel-cage induction machine rotates slower than the synchronous speed, the machine operates as a motor because the interaction of the magnetic fields in the machine creates a force that tends to increase the rotor speed. Conversely, when the rotor turns at a speed higher than the synchronous speed, the interaction of the magnetic fields creates a force that tends to slow down the motor, and thus, the machine operates as an asynchronous generator. Figure 4-13 illustrates the two cases.

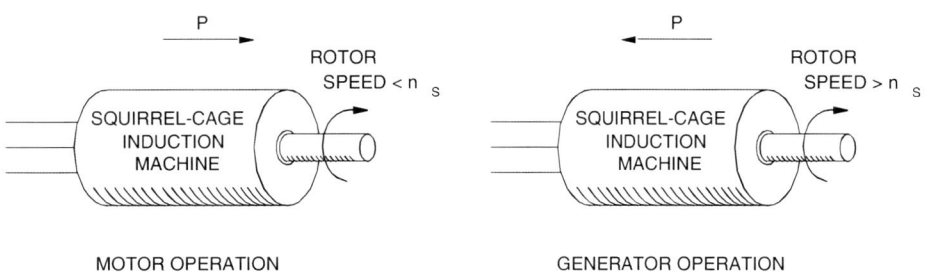

Figure 4-13. Motor or generator operation depends on the squirrel-cage rotor speed.

A particularity of the squirrel-cage induction machine is that it always requires reactive power to operate. The reactive power is needed to create the rotating magnetic field that is essential whether the machine operates as a motor or a generator. If the rotor of a squirrel-cage motor is turned without the motor being connected to an ac source, no output voltage is generated. This is because no induced current flows in the rotor. In order for the squirrel-cage induction machine to operate as an asynchronous generator, it must be connected to an ac source to obtain the reactive power necessary for the rotating magnetic field. The speed versus torque characteristic shown in Figure 4-14 illustrates both motor and generator operation of a squirrel-cage induction machine.

Ex. 4-2 – Eddy-Current Brake and Asynchronous Generator ♦ Procedure

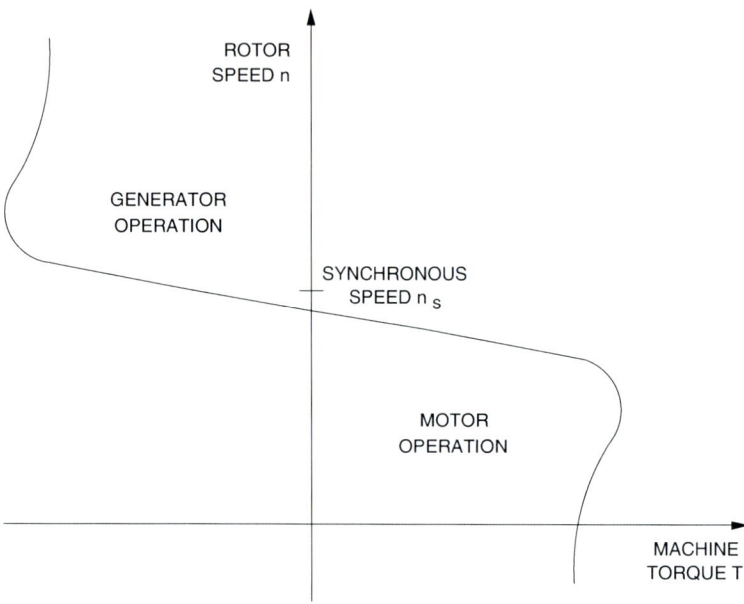

Figure 4-14. Speed versus torque characteristic of a squirrel-cage induction machine.

Procedure summary

In the first part of the exercise, you will set up the equipment in the Workstation, connect the equipment as shown in Figure 4-15, and make the appropriate settings on the equipment.

In the second part of the exercise, you will demonstrate eddy-current braking. An eddy-current brake will be implemented by connecting one of the stator winding of the Four-Pole Squirrel-Cage Induction Motor to a dc power source through a resistive load. Varying the value of the resistive load will allow variation of the electromagnet current, and thereby, variation of the braking torque.

In the third part of the exercise, you will observe the operation of a squirrel-cage induction motor operating as an asynchronous generator.

EQUIPMENT REQUIRED

Refer to the Equipment Utilization Chart in Appendix C to obtain the list of equipment required for this exercise.

PROCEDURE

 High voltages are present in this laboratory exercise. Do not make or modify any banana jack connections with the power on unless otherwise specified.

Setting up the equipment

1. Install the equipment required in the EMS workstation.

   Mechanically couple the prime mover/dynamometer module to the Four-Pole Squirrel-Cage Induction Motor.

2. On the Power Supply, make sure the main power switch is set to the O (off) position, and the voltage control knob is turned fully counterclockwise. Ensure the Power Supply is connected to a three-phase power source.

    If you are using the Four-Quadrant Dynamometer/Power Supply, Model 8960-2, connect its POWER INPUT to a wall receptacle.

3. Ensure that the data acquisition module is connected to a USB port of the computer.

   Connect the POWER INPUT of the data acquisition module to the 24 V - AC output of the Power Supply.

    If you are using the Prime Mover/Dynamometer, Model 8960-1, connect its LOW POWER INPUT to the 24 V - AC output of the Power Supply.

   On the Power Supply, set the 24 V - AC power switch to the I (on) position

    If you are using the Four-Quadrant Dynamometer/Power Supply, Model 8960-2, turn it on by setting its POWER INPUT switch to the I (on) position. Press and hold the FUNCTION button 3 seconds to have uncorrected torque values on the display of the Four-Quadrant Dynamometer/Power Supply. The indication "NC" appears next to the function name on the display to indicate that the torque values are uncorrected.

4. Start the Data Acquisition software (LVDAC or LVDAM). Open setup configuration file ACMOTOR1.DAI.

   In the Metering window, select layout 2. Make sure that the continuous refresh mode is selected.

5. Connect the equipment as shown in Figure 4-15. Connect the three resistor sections on the Resistive Load module in parallel to implement resistor R_1. Note that the resistance value of R_1 is initially set to infinite (∞).

6. Set the Four-Quadrant Dynamometer/Power Supply or the Prime Mover/Dynamometer to operate as a clockwise prime mover. To do this, refer to Exercise 1-1 or Exercise 1-2 if necessary.

    If you are performing the exercise using LVSIM®-EMS, you can zoom in the Prime Mover/Dynamometer module before setting the controls in order to see additional front panel markings related to these controls.

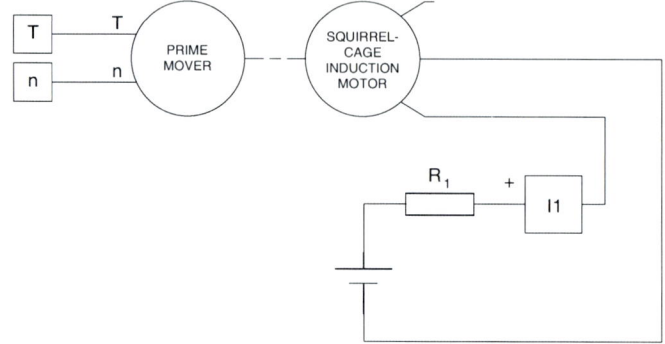

Local ac power network		R_1 (Ω)
Voltage (V)	Frequency (Hz)	
120	60	∞
220	50	∞
220	60	∞
240	50	∞

Figure 4-15. Circuit used to demonstrate eddy-current braking.

Demonstrating eddy-current braking

7. Turn the Power Supply on. Adjust the prime mover speed (indicated by the Speed meter) to 150 r/min.

   In the Metering window, set meter I line 1 in the dc mode and make sure that the torque correction function of the Torque meter is enabled. The Torque meter now indicates the braking torque $T_{BRAKING}$ caused by the squirrel-cage induction motor.

   Record the prime mover speed n, the electromagnet current I_{EM}, and the braking torque $T_{Braking}$, in the following blank spaces. These parameters are indicated by meters Speed, I line 1, and Torque, respectively. Also, record the direction of rotation.

   $n = $ _____ r/min

   $I_{EM} = $ _____ A

   $T_{BRAKING} = $ _____ N·m (lbf·in)

   Direction of rotation: _____

8. Close the switches on the Resistive Load module one at a time to increase the current in the stator electromagnet by steps. While doing this, observe the speed and torque indicated in the Metering window.

   When all switches are closed, record the prime mover speed n, the electromagnet current I_{EM}, the braking torque $T_{BRAKING}$, and the direction of rotation in the following blank spaces.

   $n = $ _____ r/min

   $I_{EM} = $ _____ A

   $T_{BRAKING} = $ _____ N·m (lbf·in)

   Direction of rotation: _____

   Turn the Power Supply off.

   Describe how the braking torque varies when the electromagnet current is increased.

   _____

   _____

   Do the results demonstrate that the squirrel-cage induction motor operate as an eddy-current brake?

   ❑ Yes   ❑ No

9. On the Resistive Load module, make the appropriate settings so that the resistance value of resistor R_1 is infinite.

   Set the Four-Quadrant Dynamometer/Power Supply or the Prime Mover Dynamometer to operate as a counterclockwise prime mover. To do this, refer to Exercise 1-1 or Exercise 1-2 if necessary.

   Turn the Power Supply on and set the prime mover speed to -150 r/min.

   Record the prime mover speed n, the electromagnet current I_{EM}, the braking torque $T_{BRAKING}$ (indicated in the Metering window) and the direction of rotation in the following blank spaces.

Ex. 4-2 – Eddy-Current Brake and Asynchronous Generator ♦ *Procedure*

$n = $ _____ r/min

$I_{EM} = $ _____ A

$T_{BRAKING} = $ _____ N·m (lbf·in)

Direction of rotation: _____

10. Close the switches on the Resistive Load module one at a time to increase the current in the stator electromagnet by steps. While doing this, observe the speed and torque indicated in the Metering window.

    When all switches are closed, record the prime mover speed n, the electromagnet current I_{EM}, the braking torque $T_{BRAKING}$, and the direction of rotation in the following blank spaces.

    $n = $ _____ r/min

    $I_{EM} = $ _____ A

    $T_{BRAKING} = $ _____ N·m (lbf·in)

    Direction of rotation: _____

    Set the prime mover speed to 0, then turn the Power Supply off.

    Describe how the braking torque varies when the electromagnet current is increased.

    _____

    _____

    Is the operation of the squirrel-cage induction motor affected by the direction of rotation of the Prime Mover?

    ❑ Yes   ❑ No

Asynchronous generator operation

11. Modify the connections so that the equipment is connected as shown in Figure 4-16. Do not connect lines A, B, and C of the three-phase power source to the circuit for now. This will be done later in the exercise.

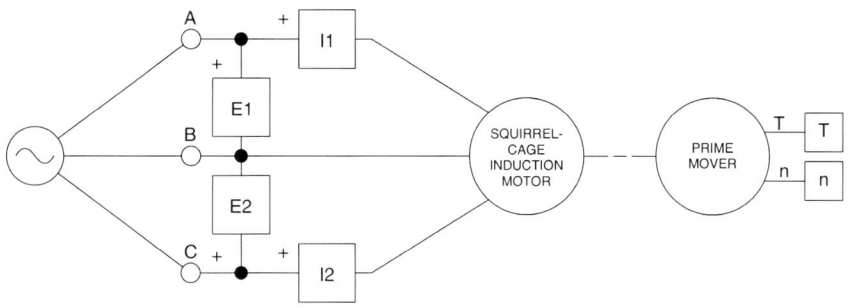

Figure 4-16. Circuit used to demonstrate asynchronous generator operation.

Go to the next step if you are using the Four-Quadrant Dynamometer/Power Supply, Model 8960-2. Otherwise, turn the Power Supply on then adjust the voltage control knob so that the prime mover speed (indicated by the speed meter) is equal to the synchronous speed of the Four-Pole Squirrel-Cage Induction Motor. If the prime mover rotates counterclockwise, turn the Power Supply off, reverse the connection of the leads at the PRIME MOVER INPUT of the Prime Mover/Dynamometer, turn the Power Supply on, and readjust the prime mover speed (if necessary).

Turn the Power Supply off. Do not modify the setting of the voltage control knob (i.e., the prime mover speed setting).

12. Connect lines A, B, and C of the three-phase power source to the circuit as shown in Figure 4-16.

    Set the Four-Quadrant Dynamometer/Power Supply or the Prime Mover/Dynamometer to operate as brake.

     If you are using the Prime Mover/Dynamometer, Model 8960-1, set the MODE switch to the DYNamometer position to do so. Also, leave the PRIME MOVER INPUT connected to the Power Supply.

    Set the brake torque control to minimum (fully counterclockwise position).

    In the Metering window, set meter I line 1 in the ac mode and set meters Act. power and Mech. power so that they have a zero-centre analog-type display and a 750-W scale.

13. Turn the Power Supply on and verify that the squirrel-cage induction machine rotates clockwise.

     In the squirrel-cage induction machine rotates in the counterclockwise direction, turn the Power Supply off, interchange two of the three leads supplying power to the machine, and turn the Power Supply on.

14. On the brake, set the torque control so that the torque indicated by the Torque meter in the Metering window is about 1.3 N·m (about 11.5 lbf·in).

Record in the following blank spaces the squirrel-cage induction machine active power P, reactive power Q, mechanical power P_m, speed n, and torque T. These parameters are indicated by meters Act. power, React. power, Mech. power, Speed, and Torque, respectively.

$P = $ _____ W

$Q = $ _____ var

$P_m = $ _____ W

$n = $ _____ r/min

$T = $ _____ N · m (lbf · in)

Does active power flow from the ac power source to the squirrel-cage induction machine?

❏ Yes   ❏ No

What does this indicate about the operation of the squirrel-cage induction machine?

_____

_____

15. On the brake, set the torque control to minimum (fully counterclockwise position).

Set the Four-Quadrant Dynamometer/Power Supply or the Prime Mover/Dynamometer to operate as prime mover rotating at the synchronous speed of the Four-Pole Squirrel-Cage Induction Motor **(read the notes below before performing this manipulation)**.

 If you are using the Prime Mover/Dynamometer, Model 8960-1, set the MODE switch to the PRIME MOVER position to switch to prime mover operation.

 If you are using the Four-Quadrant Dynamometer/Power Supply, Model 8960-2, proceed as follow to switch to prime mover operation: stop the Two-Quadrant, Constant-Torque Brake, select the CW Prime Mover/Brake function, set the Speed parameter to the synchronous speed of the Four-Pole Squirrel-Cage Induction Motor, and start the CW Prime Mover/Brake.

Readjust the prime mover speed so that the machines rotate at the synchronous speed of the Four-Pole Squirrel-Cage Induction Motor.

Record in the following blank spaces the squirrel-cage induction machine active power P, reactive power Q, mechanical power P_m, speed n, and torque T. These parameters are indicated by meters Act. power, React. power, Mech. power, Speed, and Torque, respectively.

$P =$ _____ W

$Q =$ _____ var

$P_m =$ _____ W

$n =$ _____ r/min

$T =$ _____ N·m (lbf·in)

Does a significant amount of active power flow between the ac power source and the squirrel-cage induction machine?

❏ Yes   ❏ No

16. Slowly adjust the prime mover speed so that the machines rotate at 105% of the synchronous speed of the Four-Pole Squirrel-Cage Induction Motor.

Record in the following blank spaces the squirrel-cage induction machine active power P, reactive power Q, mechanical power P_m, speed n, and torque T.

$P =$ _____ W

$Q =$ _____ var

$P_m =$ _____ W

$n =$ _____ r/min

$T =$ _____ N·m (lbf·in)

Does active power flow from the squirrel-cage induction machine to the ac power source?

❏ Yes   ❏ No

What does this indicate about the operation of the squirrel-cage induction machine?

17. Turn the Power Supply off.

Disconnect the three-phase power source from the circuit at points A, B, and C shown in Figure 4-16.

Turn the Power Supply on and readjust the prime mover speed so that the machine rotates at 105% of the synchronous speed of the Four-Pole Squirrel-Cage Induction Motor.

Record the line voltage E_{LINE} generated by the asynchronous generator (indicated by meter E line 1).

$E_{LINE} = $ _____ V

Does this confirm that generator operation is not possible unless the squirrel-cage induction machine is connected to a three-phase ac power network?

❏ Yes   ❏ No

18. Turn all equipment off.

Remove all leads and cables.

Additional experiments

Speed versus torque characteristic of a squirrel-cage induction motor for both the motor and generator modes of operation

You can obtain the speed versus torque characteristic of a squirrel-cage induction motor for both the motor and generator modes of operation. To do so, set the Four-Quadrant Dynamometer/Power Supply or the Prime Mover/Dynamometer to operate as a clockwise prime mover. (If you are using the Four-Quadrant Dynamometer/Power Supply, make sure to press and hold the FUNCTION button 3 seconds to have uncorrected torque values on this module's display). Open setup configuration file ACMOTOR1.DAI and select meter layout 2. Make sure the torque correction function of the Torque meter is enabled. Make sure the Power Supply is turned off and refer to steps 11 to 13 of this exercise to set up the circuit shown in Figure 4-16. Clear the data recorded in the Data Table (if any). Set the Four-Quadrant Dynamometer/Power Supply or the Prime Mover/Dynamometer to operate as a prime mover and adjust the prime mover speed until the Torque meter indicates -2.7 N·m (-24.0 lbf·in). The machine speed should be greater than the synchronous speed when the torque is -2.7 N·m (-24.0 lbf·in). Adjust the prime mover speed so that the torque indicated by the Torque meter passes from -2.7 N·m (-24.0 lbf·in) to 0.0 N·m (0.0 lbf·in) in steps of 0.3 N·m (3.0 lbf·in). For each torque setting record the data in the Data Table.

Ex. 4-2 – Eddy-Current Brake and Asynchronous Generator ♦ *Conclusion*

After recording the data for 0.0 N·m (0.0 lbf·in), set the Four-Quadrant Dynamometer/Power Supply or the Prime Mover/Dynamometer to operate as a brake, then make sure the brake torque control on the brake is set to minimum (turned fully counterclockwise). Refer to step 9 of Exercise 4-1 to complete the measurements. Turn the equipment off. Edit the Data Table so as to keep only the values of the line voltage E_{LINE}, line current I_{LINE}, active power P, reactive power Q, speed n, and torque T. Entitle the Data Table as DT421. Plot a graph of the speed (obtained from the Speed meter) as a function of the torque (obtained from the Torque meter). Entitle the graph as G421.

CONCLUSION

In this exercise, you demonstrated the concept of eddy-current braking using a squirrel-cage induction motor. You observed that the braking torque increases as the dc current flowing in the stator electromagnet increases. You observed that a three-phase squirrel-cage induction motor can operate as an asynchronous generator when it rotates at a speed higher than the synchronous speed n_S. You saw that the squirrel-cage induction motor returns active power to the ac power network when it operates as a generator. You observed that reactive power is always required by the squirrel-cage induction motor to create the rotating magnetic field, whether it operates as a motor or generator.

If you have performed the additional experiments, you plotted a graph of speed versus torque for a squirrel-cage induction motor. This graph covers both the motor and generator modes of operation.

REVIEW QUESTIONS

1. A fixed electromagnet that creates braking torque which acts on a squirrel-cage rotor describes a magnetic

    a. field.
    b. wire.
    c. brake (eddy-current brake).
    d. flux.

2. When a squirrel-cage induction motor turns faster than the synchronous speed determined by the ac network, it

    a. consumes active and reactive power.
    b. consumes active power and supplies reactive power.
    c. supplies active power and supplies reactive power.
    d. supplies active power and consumes reactive power.

3. A squirrel-cage induction motor always requires

    a. reactive power to create the rotating magnetic field.
    b. active power to create the rotating magnetic field.
    c. a source of dc power to operate correctly.
    d. a prime mover to help it start.

4. The speed of a squirrel-cage induction motor operating as an asynchronous generator

   a. is less than the synchronous speed.
   b. is equal to the synchronous speed.
   c. is greater than the synchronous speed.
   d. depends on the direction of rotation.

5. What will be the output voltage of an asynchronous generator turning at synchronous speed if the generator is not connected to an ac source?

   a. It will be zero except for a small voltage due to residual magnetism.
   b. It will depend on the direction of rotation.
   c. It will equal the nominal voltage rating of the generator.
   d. It will be much higher than if the generator were connected to an ac source.

Exercise 4-3

Effect of Voltage on the Characteristics of Induction Motors

EXERCISE OBJECTIVE

When you have completed this exercise, you will be able to use the Four-Pole Squirrel-Cage Induction Motor module to demonstrate how the voltage applied to an induction motor affects its characteristics.

DISCUSSION

It is desirable to have a strong rotating magnetic field in induction motors to obtain the strongest magnetic force possible between the stator and the rotor. This results in a powerful motor because this allows a high torque to be developed. To increase the strength of the rotating magnetic field, it is necessary to increase the ac voltage applied to the stator windings of the induction motor (motor voltage). However, when the motor voltage is increased too much, the motor current (current in the stator windings) is large even at no load because the iron in the stator of the motor begins to saturate. When the motor is saturated, the strength of the rotating magnetic field almost ceases to increase as the no-load motor current is increased. To determine the nominal voltage of an induction motor, a voltage versus current graph as shown in Figure 4-17 is usually plotted when the motor operates without load. This graph is similar to the saturation curve of a transformer or dc motor. The nominal voltage is selected so that the motor operating point is located near or in the knee of the saturation curve.

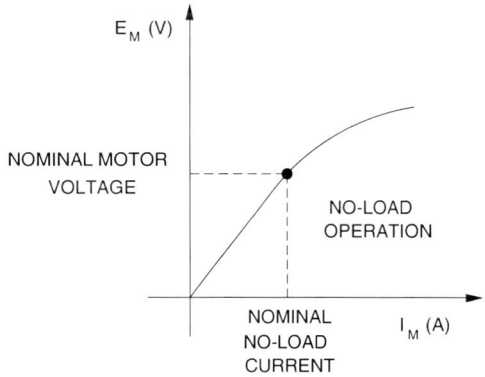

Figure 4-17. No-load voltage versus current characteristic of an induction motor.

It is also possible to plot the speed versus torque characteristic for different motor voltages. Figure 4-18 shows an example of speed versus torque characteristics for both nominal and reduced motor voltages.

Ex. 4-3 – Effect of Voltage on the Characteristics of Induction Motors ♦ *Discussion*

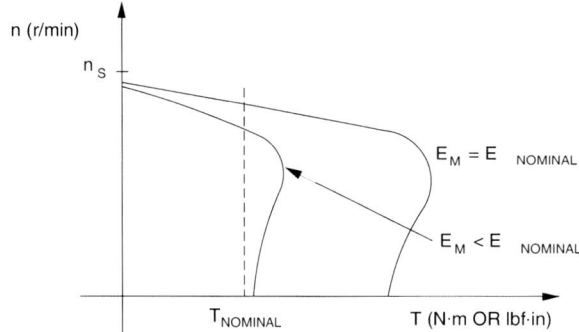

Figure 4-18. Speed versus torque characteristics for nominal and reduced motor voltages.

As shown in Figure 4-18, both the locked-rotor torque and the breakdown torque decrease greatly when the motor voltage is reduced. In practice, the torque decreases by a factor equal to the square of the reduction factor of the motor voltage. For example, the torque is reduced by a factor of four when the motor voltage is reduced by a factor of two (i.e., decreased to one half its original value). In some circumstances, the motor voltage is reduced intentionally to obtain small variations in the speed of an induction motor. Furthermore, reducing the motor voltage allows the starting current of the motor to be lowered.

Procedure summary

In the first part of the exercise, you will set up the equipment in the Workstation and connect the equipment as shown in Figure 4-19.

In the second part of the exercise, you will vary the voltage applied to the windings of the squirrel-cage induction motor at no load while measuring and recording the winding current. You will plot a graph of the winding voltage versus the winding current and observe the effect of saturation.

In the third part of the exercise, you will set up the circuit shown in Figure 4-20 and make the appropriate settings on the equipment. You will then set the voltage applied to the squirrel-cage induction motor below the nominal value to see the effect this has on the no-load speed.

In the fourth part of the exercise, you will vary the load applied to the squirrel-cage induction motor operating with reduced voltage. For each load setting, you will record in the Data Table various electrical and mechanical parameters related to the motor. You will then use this data to plot various graphs and determine many of the characteristics of the squirrel-cage induction motor when it operates with reduced voltage.

EQUIPMENT REQUIRED

Refer to the Equipment Utilization Chart in Appendix C to obtain the list of equipment required for this exercise.

Ex. 4-3 – Effect of Voltage on the Characteristics of Induction Motors ♦ Procedure

PROCEDURE

⚠ WARNING

High voltages are present in this laboratory exercise. Do not make or modify any banana jack connections with the power on unless otherwise specified.

Setting up the equipment

1. Install the equipment required in the EMS workstation.

2. On the Power Supply, make sure the main power switch is set to the O (off) position, and the voltage control knob is turned fully counterclockwise. Ensure the Power Supply is connected to a three-phase power source.

    If you are using the Four-Quadrant Dynamometer/Power Supply, Model 8960-2, connect its POWER INPUT to a wall receptacle.

3. Ensure that the data acquisition module is connected to a USB port of the computer.

   Connect the POWER INPUT of the data acquisition module to the 24 V - AC output of the Power Supply.

    If you are using the Prime Mover/Dynamometer, Model 8960-1, connect its LOW POWER INPUT to the 24 V - AC output of the Power Supply.

   On the Power Supply, set the 24 V - AC power switch to the I (on) position.

    If you are using the Four-Quadrant Dynamometer/Power Supply, Model 8960-2, turn it on by setting its POWER INPUT switch to the I (on) position. Press and hold the FUNCTION button 3 seconds to have uncorrected torque values on the display of the Four-Quadrant Dynamometer/Power Supply. The indication "NC" appears next to the function name on the display to indicate that the torque values are uncorrected.

4. Start the Data Acquisition software (LVDAC or LVDAM). Open setup configuration file ACMOTOR1.DAI.

    If you are using LVSIM-EMS in LVVL, you must use the IMPORT option in the File menu to open the configuration file.

   In the Metering window, select layout 2. Make sure that the continuous refresh mode is selected.

5. Connect the equipment as shown in Figure 4-19.

    The windings of the Four-Pole Squirrel-Cage Induction Motor are connected in delta to allow a greater voltage to be applied to the windings.

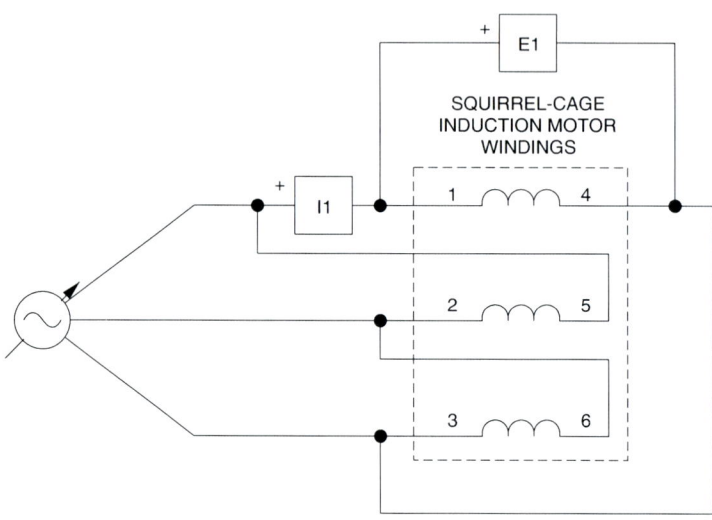

Figure 4-19. Delta connection of the stator windings of the four-pole squirrel-cage induction motor.

Induction motor saturation

6. Turn the Power Supply on and set the voltage control knob so that the voltage applied to each of the squirrel-cage induction motor windings (indicated by meter E line 1) is equal to 50% of the nominal voltage of these windings.

    The nominal voltage and current of the windings of the Four-Pole Squirrel-Cage Induction Motor are indicated by the module front panel.

   Record the winding voltage and the winding current (indicated by meter I line 1) in the Data Table.

7. On the Power Supply, turn the voltage control knob in 5% increments up to the 100% position to increase the winding voltage by steps. For each voltage setting, record the winding voltage and current in the Data Table.

    The nominal line current of the Four-Pole Squirrel-Cage Induction Motor is exceeded while performing this manipulation. It is, therefore, suggested to complete the manipulation within a time interval of 5 minutes or less.

   When all data has been recorded, turn the voltage control knob fully counterclockwise and turn the Power Supply off.

8. In the Data Table window, confirm that the data has been stored, entitle the Data Table as DT431, and print the Data Table

    Refer to the user guide dealing with the computer-based instruments for EMS to know how to edit, entitle, and print a Data Table.

9. In the Graph window, make the appropriate settings to obtain a graph of the motor winding voltage (obtained from meter E line 1) as a function of the motor winding current (obtained from meter I line 1). Entitle the graph as G431, name the x-axis as Squirrel-cage induction-motor winding current, name the y-axis as Squirrel-cage induction-motor winding voltage, and print the graph.

Refer to the user guide dealing with the computer-based instruments for EMS to know how to use the Graph window of the Metering application to obtain a graph, entitle a graph, name the axes of a graph, and print a graph.

10. Indicate the nominal winding voltage of the squirrel-cage induction motor in graph G431.

    Is the nominal winding voltage located near the knee of the motor saturation curve?

    ❏ Yes ❏ No

    Referring to graph G431, does the equivalent impedance of the induction motor at no-load seem to decrease as the winding voltage is increased?

    ❏ Yes ❏ No

11. Use graph G431 to approximate the winding voltage ($E_{WINDING}$) at which nominal current flows in the motor windings (when no load is applied to the motor).

    $E_{WINDING}$: _____ V (at nominal winding current and with no load)

    If the motor is operated at this voltage, winding current will exceed the nominal value as soon as the motor is mechanically loaded and the motor will overheat.

Effect of voltage on the speed of an induction motor

12. Remove all leads except the 24-V ac power cables.

    Mechanically couple the prime mover/dynamometer module to the Four-Pole Squirrel-Cage Induction Motor.

    Connect the equipment as shown in Figure 4-20.

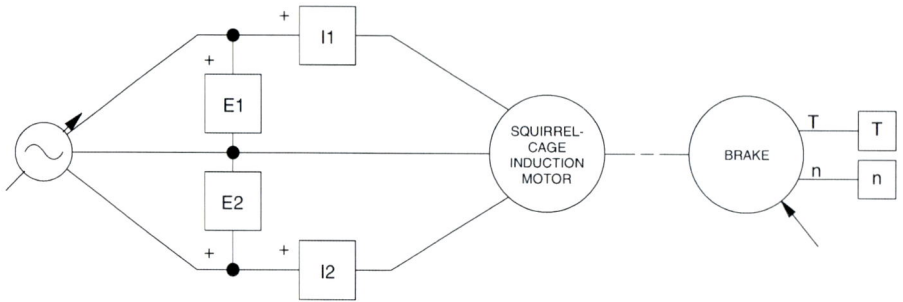

Figure 4-20. Squirrel-cage induction motor coupled to a brake.

13. Set the Four-Quadrant Dynamometer/Power Supply or the Prime Mover/Dynamometer to operate as a brake, then set the brake torque control to minimum (fully counterclockwise position). To do this, refer to Exercise 1-1 or Exercise 1-2 if necessary.

 If you are performing the exercise using LVSIM®-EMS, you can zoom in the Prime Mover/Dynamometer module before setting the controls in order to see additional front panel markings related to these controls.

14. Turn the Power Supply on and set the voltage control knob so that the line voltage indicated by meter E line 1 is equal to 75% of the nominal line voltage of the squirrel-cage induction motor.

 The rating of any of the machines is indicated in the lower left corner of the module front panel.

Record in the following blank space the no-load speed n of the motor indicated by the Speed meter in the Metering window.

$n =$ _____ r/min (at 75% of the nominal motor line voltage)

Is the no-load speed obtained when the line voltage is set to 75% of the nominal value less than the no-load speed obtained when the line voltage is set to the nominal value, as in step 7 of Exercise 4-1?

❏ Yes ❏ No

Does changing the voltage applied to the squirrel-cage induction motor allow the speed to be changed?

❏ Yes ❏ No

Induction motor characteristics at reduced voltage

15. (If you are using the Four-Quadrant Dynamometer/Power Supply, make sure to press and hold the FUNCTION button 3 seconds to have uncorrected torque values on this module's display). In the Metering window, clear the data recorded in the Data Table and make sure that the torque correction function of the Torque meter is enabled. The Torque meter now indicates the output torque of the squirrel-cage induction motor.

Record the motor line voltage E_{LINE}, line current I_{LINE}, active power P, reactive power Q, output torque T, and speed n in the Data Table. These parameters are indicated by meters E line 1, I line 1, Act. power, React. power, Torque, and Speed, respectively.

On the brake, set the torque control so that the torque indicated by the Torque meter in the Metering window increases by 0.3 N·m (2.0 lbf·in) increments up to 1.2 N·m (11.0 lbf·in). For each torque setting, record the data in the Data Table.

On the brake, continue to adjust the torque control so that the torque indicated by the Torque meter on the module display increases by 0.1 N·m (1.0 lbf·in) increments, until the motor speed starts to decrease fairly rapidly (breakdown torque region). For each torque setting, record the data in the Data Table.

Once the motor speed has stabilized, record the data in the Data Table.

 The nominal line current of the Four-Pole Squirrel-Cage Induction Motor may be exceeded while performing this manipulation. It is, therefore, suggested to complete the manipulation within a time interval of 5 minutes or less.

16. When all data has been recorded, set the torque control on the brake to minimum (fully counterclockwise), turn the voltage control knob fully counterclockwise, and turn the Power Supply off.

In the Data Table window, confirm that the data has been stored, entitle the Data Table as DT432, and print the Data Table.

Does the motor line current indicated in column I line 1 increase as the mechanical load applied to the squirrel-cage induction motor increases?

❑ Yes   ❑ No

Ex. 4-3 – Effect of Voltage on the Characteristics of Induction Motors ♦ *Procedure*

17. In the Graph window, make the appropriate settings to obtain a graph of the motor speed (obtained from meter N) as a function of the motor torque (obtained from meter T). Entitle the graph as G432, name the x-axis as Squirrel-cage induction-motor torque, name the y-axis as Squirrel-cage induction-motor speed, and print the graph.

    Determine the breakdown torque of the squirrel-cage induction motor using graph G432.

    $T_{BREAKDOWN}$ = _____ N · m (lbf · in)

    (with motor voltage reduced to 75% of the nominal value)

    Determine the minimum-speed torque using graph G432. This torque is a good approximation of the locked-rotor torque of the squirrel-cage induction motor.

    $T_{LOCKED\ ROTOR}$ = _____ N · m (lbf · in)

    (with motor voltage reduced to 75% of the nominal value)

    Compare the breakdown torque and locked-rotor torque obtained when the motor voltage is set to 75% of the nominal value to those obtained when the motor voltage is set to the nominal value, as in step 12 of Exercise 4-1.

    _____

    _____

    _____

    Does reducing the motor voltage decrease the torque developed by the motor?

    ❑ Yes   ❑ No

18. In the Graph window, make the appropriate settings to obtain a graph of the motor active (P) and reactive (Q) power values (obtained from meters Act. power and React. power, respectively) as a function of the motor speed (obtained from the Speed meter) using the data recorded previously in Data Table DT432. Entitle the graph as G432-1, name the x-axis as Squirrel-cage induction-motor speed, name the y-axis as Squirrel-cage induction-motor active power and reactive power, and print the graph.

Ex. 4-3 – Effect of Voltage on the Characteristics of Induction Motors ♦ *Procedure*

Compare the active and reactive power values obtained when the motor voltage is set to 75% of the nominal value (graph G432-1) to those obtained when the motor voltage is set to the nominal value (graph G411-1 obtained in Exercise 4-1).

_____

_____

_____

19. In the Graph window, make the appropriate settings to obtain a graph of the motor line current I_{LINE} (obtained from meter I line 1) as a function of the motor speed (obtained from the Speed meter), using the data recorded previously in Data Table DT432. Entitle the graph as G432-2, name the x-axis as Squirrel-cage induction-motor speed, name the y-axis as Squirrel-cage induction-motor line current, and print the graph.

Compare the starting current (line current at low speeds) obtained when the motor voltage is set to 75% of the nominal value (graph G432-2) to that obtained when the motor voltage is set to the nominal value (graph G411-2 obtained in Exercise 4-1).

_____

_____

_____

Does reducing the motor voltage decrease the motor starting current?

❑ Yes   ❑ No

20. On the Power Supply, set the 24 V - AC power switch to the O (off) position.

 If you are using the Four-Quadrant Dynamometer/Power Supply, Model 8960-2, turn it off by setting its POWER INPUT switch to the O (off) position.

Remove all leads and cables.

Ex. 4-3 – Effect of Voltage on the Characteristics of Induction Motors ♦ *Conclusion*

CONCLUSION

In this exercise, you observed that the winding current increases greatly when the nominal winding voltage is exceeded because saturation occurs in the squirrel-cage induction motor. You plotted the saturation curve of the squirrel-cage induction motor and found that the nominal voltage of the motor is located near the knee of the curve. You plotted a graph of speed versus torque with reduced voltage applied to the squirrel-cage induction motor. You used this graph to determine the breakdown and locked-rotor torques of the motor. You found that reducing the motor voltage decreases the torque developed by the motor at any speed. You also plotted a graph of the motor active and reactive powers versus speed and observed that the squirrel-cage induction motor draws less power from the ac power source when the motor voltage is reduced. Finally, you plotted a graph of the motor line current versus speed and observed that reducing the motor voltage decreases the starting current (line current at low speeds).

REVIEW QUESTIONS

1. How is motor torque affected when the motor voltage is decreased?

    a. It decreases.
    b. It increases.
    c. It does not change.
    d. It depends on the speed of the motor.

2. What variation in torque is caused by a motor voltage reduction of 50%?

    a. An increase of 25%.
    b. A decrease of 50%.
    c. A decrease of 75%.
    d. A decrease of 100%.

3. When the strength of the stator electromagnet is increased, the torque produced by a squirrel-cage induction motor

    a. does not change.
    b. decreases.
    c. increases.
    d. Torque only depends on the size of the motor.

4. The current in the stator winding of a squirrel-cage induction motor greatly increases when the nominal winding voltage is exceeded because

    a. the motor develops a large torque.
    b. saturation occurs in the motor.
    c. squirrel-cage reaction occurs in the motor.
    d. because reactive power is consumed in the motor.

5. What advantage is obtained by reducing the voltage applied to a squirrel-cage induction motor?

   a. The line current is reduced during starting.
   b. The motor brushes suffer less damage.
   c. The starting torque is increased.
   d. The danger of motor runaway is avoided.

Exercise 4-4

Single-Phase Induction Motors

EXERCISE OBJECTIVE

When you have completed this exercise, you will be able to demonstrate the main operating characteristics of single-phase induction motors using the Capacitor-Start Motor module.

DISCUSSION

It is possible to obtain a single-phase squirrel-cage induction motor using a simple electromagnet connected to a single-phase ac power source as shown in Figure 4-21.

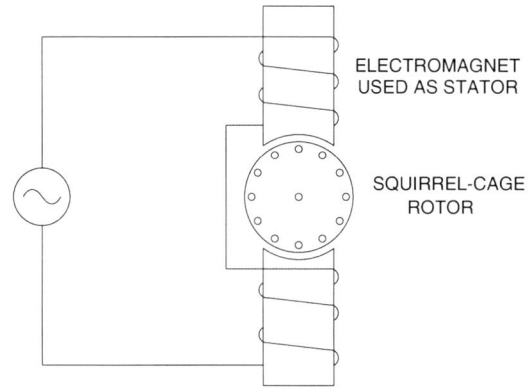

Figure 4-21. Simple single-phase squirrel-cage induction motor.

The operating principle of this type of motor is more complex than that of the three-phase squirrel-cage induction motor. The simple induction motor of Figure 4-21 can even be considered as an eddy-current brake that brakes in an intermittent manner since the sinusoidal current in the stator electromagnet continually passes from peaks to zeros. One could even wonder how this motor can turn since it seems to operate similarly as an eddy-current brake.

However, when the rotor of the simple induction motor of Figure 4-21 is turned manually, a torque which acts in the direction of rotation is produced, and the motor continues to turn as long as ac power is supplied to the stator electromagnet. This torque is due to a rotating magnetic field that results from the interaction of the magnetic field produced by the stator electromagnet and the magnetic field produced by the currents induced in the rotor. A graph of speed versus torque for this type of motor is shown in Figure 4-22. The curve shows that the torque is very small at low speeds. It increases to a maximum value as the speed increases, and finally decreases towards zero again when the speed approaches the synchronous speed n_S.

Ex. 4-4 – Single-Phase Induction Motors ♦ *Discussion*

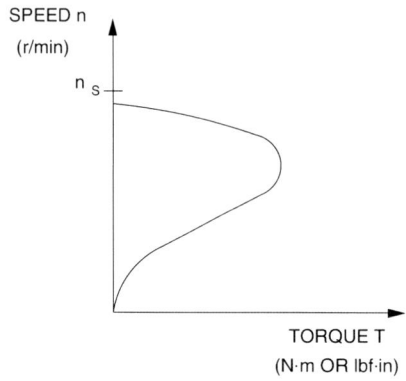

Figure 4-22. Speed versus torque characteristic of a single-phase induction motor.

The low torque values at low speeds are due to the fact that the currents induced in the rotor produce magnetic fields that create forces which act on the rotor in various directions. Most of these forces cancel each other and the resulting force acting on the rotor is weak. This explains why the single-phase induction motor shown in Figure 4-21 must be started manually. To obtain torque at low speeds (starting torque), a rotating magnetic field must be produced in the stator when the motor is starting. In Unit 1 of this manual, you saw that it is possible to create a rotating magnetic field using two alternating currents, I_1 and I_2, that are phase shifted 90° from one another, and two electromagnets placed at right angles to each other.

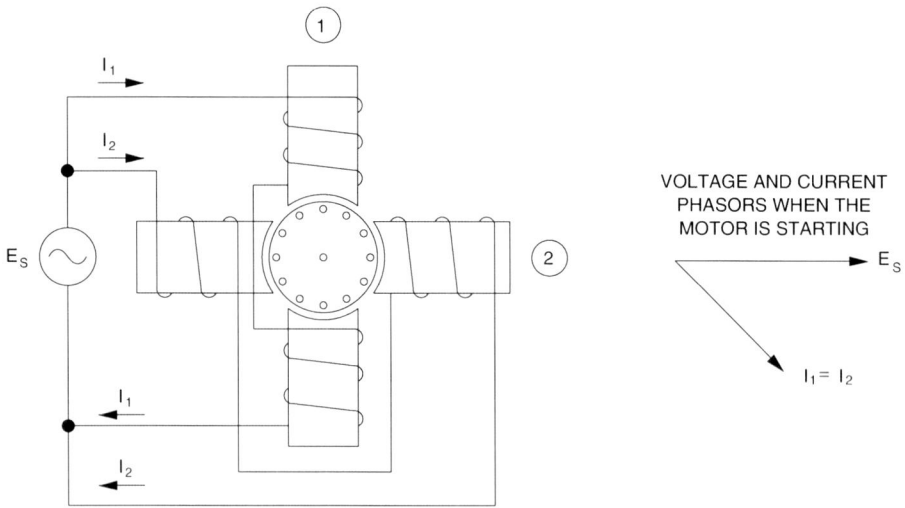

Figure 4-23. Adding a second electromagnet to the simple induction motor of Figure 4-21.

Figure 4-23 shows the simple induction motor of Figure 4-21 with the addition of a second electromagnet placed at right angle to the first electromagnet. The second electromagnet is identical to the first one and is connected to the same ac power source. The currents I_1 and I_2 in the electromagnets (winding currents) are in phase because the coils have the same impedance. However, because of the inductance of the coils of the electromagnets, there is a phase shift between the currents and the ac source voltage as illustrated in the phasor diagram of Figure 4-23.

Since currents I_1 and I_2 are in phase, there is no rotating magnetic field produced in the stator. However, it is possible to phase shift current I_2 by connecting a capacitor in series with the winding of electromagnet 2. The capacitance of the capacitor can be selected so that current I_2 leads current I_1 by 90° when the motor is starting as shown in Figure 4-24. As a result, an actual rotating magnetic field like that previously illustrated in Unit 1 is created when the motor is starting. The capacitor creates the equivalent of a two-phase ac power source and allows the motor to develop starting torque.

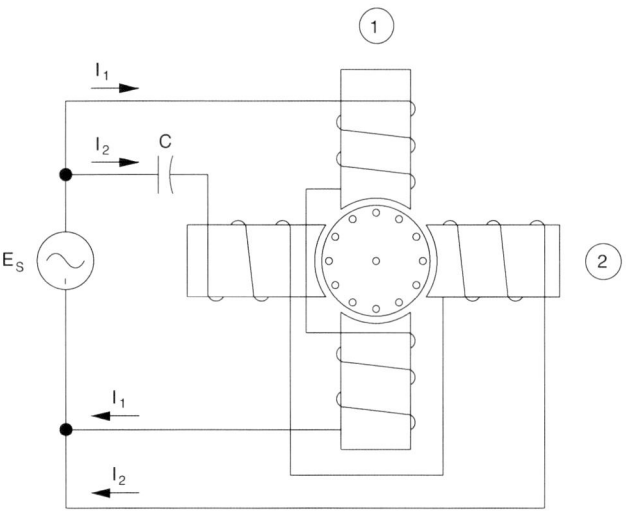

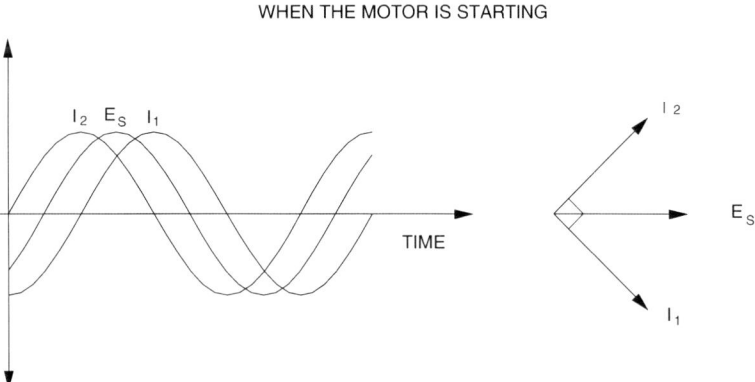

Figure 4-24. Adding a capacitor allows the induction motor to develop starting torque.

Another way to create a phase shift between currents I_1 and I_2 is to make a winding with fewer turns of smaller-sized wire. The resulting winding, which is called auxiliary winding, has more resistance and less inductance, and the winding current is almost in phase with the source voltage. Although the phase shift between the two currents is less than 90° when the motor is starting, as shown in Figure 4-25, a rotating magnetic field is created. The torque produced is sufficient for the motor to start rotating in applications not requiring high values of starting torque.

Ex. 4-4 – Single-Phase Induction Motors ♦ *Discussion*

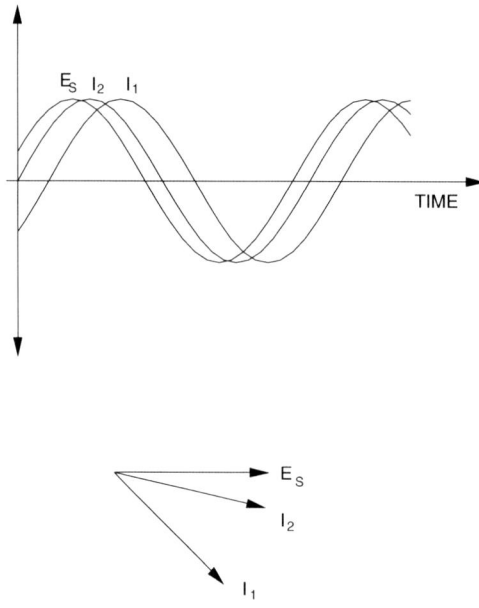

Figure 4-25. Phase shift between the winding currents when an auxiliary winding is used.

However, the auxiliary winding cannot support high currents for more than a few seconds without being damaged because it is made of fine wire. It is therefore connected through a centrifugal switch which opens and disconnects the winding from the motor circuit when the motor reaches about 75% of the normal speed. After the centrifugal switch opens, the rotating magnetic field is maintained by the interaction of the magnetic fields produced by the stator and the rotor.

Procedure summary

In the first part of the exercise, you will set up the equipment in the Workstation and connect the equipment as shown in Figure 4-26.

In the second part of the exercise, you will observe both two-phase and single-phase operation of the three-phase squirrel-cage induction motor using the Phasor Analyzer.

In the third part of the exercise, you will observe the operation of a single-phase induction motor using a capacitor-start motor and the Phasor Analyzer.

EQUIPMENT REQUIRED

Refer to the Equipment Utilization Chart in Appendix C to obtain the list of equipment required for this exercise.

Procedure

⚠ WARNING

High voltages are present in this laboratory exercise. Do not make or modify any banana jack connections with the power on unless otherwise specified.

Setting up the equipment

1. Install the equipment required in the EMS workstation.

2. On the Power Supply, make sure the main power switch is set to the O (off) position, and the voltage control knob is turned fully counterclockwise. Ensure the Power Supply is connected to a three-phase power source.

3. Ensure that the data acquisition module is connected to a USB port of the computer.

   Connect the POWER INPUT of the data acquisition module to the 24 V - AC output of the Power Supply.

   On the Power Supply, set the 24 V - AC power switch to the I (on) position.

4. Start the Data Acquisition software (LVDAC or LVDAM). Open setup configuration file ACMOTOR1.DAI.

    If you are using LVSIM-EMS in LVVL, you must use the IMPORT option in the File menu to open the configuration file.

   In the Metering window, select layout 1. Make sure that the continuous refresh mode is selected.

Ex. 4-4 – Single-Phase Induction Motors ◆ Procedure

5. Connect the equipment as shown in Figure 4-26.

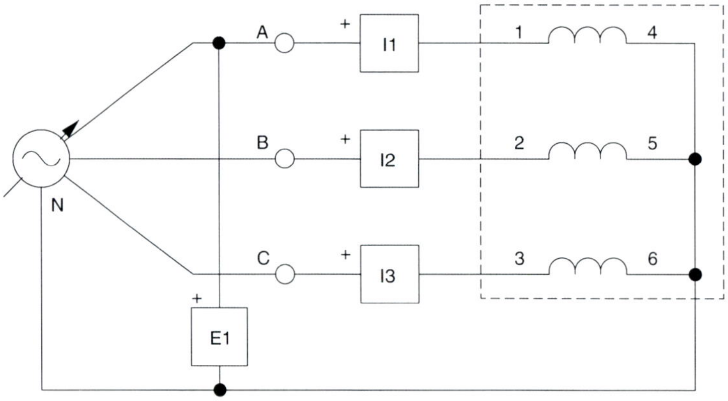

Figure 4-26. Three-phase squirrel-cage induction motor.

Two-phase and single-phase operation of a three-phase squirrel-cage induction motor

6. Turn the Power Supply on and set the voltage control knob so that the voltage applied to each of the motor windings (indicated by meter E line 1) is equal to the nominal voltage of these windings.

    The nominal voltage and current of the windings of the Four-Pole Squirrel-Cage Induction Motor are indicated on the module front panel.

    Does the squirrel-cage induction motor start readily and rotate normally?

    ❑ Yes   ❑ No

7. In the Phasor Analyzer window, select the phasor of the ac source line-to-neutral voltage (channel E1) as the reference phasor, then make the proper settings to observe this voltage, as well as the phasors representing the line currents in the three-phase squirrel-cage induction motor (channels I1, I2, and I3).

    Are the line current phasors (I1, I2, and I3) all equal in magnitude and separated by a phase angle of 120°, thus showing they create a normal rotating magnetic field?

    ❑ Yes   ❑ No

8. Turn the Power Supply off.

    Open the circuit at point A shown in Figure 4-26. Make sure that input VOLTAGE E1 of the data acquisition module remains connected to the ac power source.

9. Turn the Power Supply on.

   Does the squirrel-cage induction motor start readily and rotate normally?

   ❏ Yes   ❏ No

   In the Phasor Analyzer window, observe the current phasors on channels I2 and I3. Is there a phase shift between these phasors to create a rotating magnetic field?

   ❏ Yes   ❏ No

10. Turn the Power Supply off and turn the voltage control knob fully counterclockwise.

    Open the circuit at point B shown in Figure 4-26.

11. Turn the Power Supply on, set the voltage control knob to about 50%, wait approximately 5 seconds, then turn the Power Supply off and turn the voltage control knob fully counterclockwise.

    Does the squirrel-cage induction motor start readily and rotate normally?

    ❏ Yes   ❏ No

12. Use the Capacitive Load module to connect a capacitor (C_1) to the motor circuit as shown in Figure 4-27. Set the capacitance of the capacitor to the value indicated in the figure.

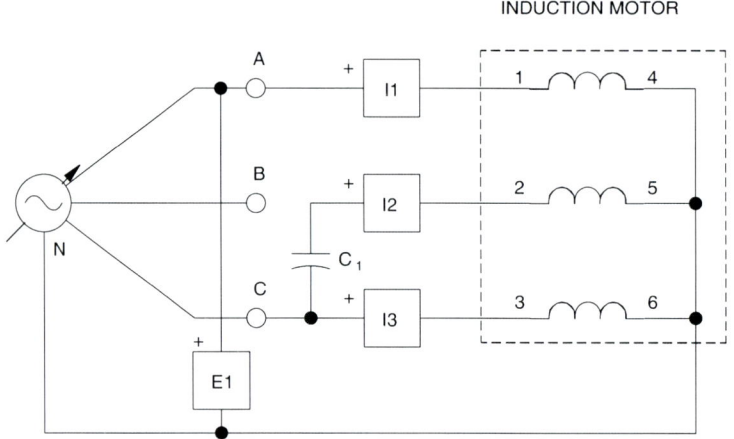

Local ac power network		C_1 (μF)
Voltage (V)	Frequency (Hz)	
120	60	15.4
220	50	5.1
220	60	4.2
240	50	4.6

Figure 4-27. Adding a capacitor to the motor circuit.

13. Turn the Power Supply on and slowly set the voltage control knob to 100%. While doing this, observe phasors I2 and I3 in the Phasor Analyzer window as the voltage increases.

Does the squirrel-cage induction motor start to rotate? Briefly explain why.

_____
_____
_____

14. On the Capacitive Load module, open the switches to disconnect the capacitor from the motor circuit and cut off the current in one of the two windings of the squirrel-cage induction motor.

Does the squirrel-cage induction motor continue to rotate, thus showing that it can operate on single-phase ac power once it has started?

❏ Yes   ❏ No

Turn the Power Supply off and turn the voltage control knob fully counterclockwise.

Operation of a single-phase induction motor (capacitor-start type)

15. Remove all leads except the 24-V ac power cable then set up the capacitor-start motor circuit shown in Figure 4-28.

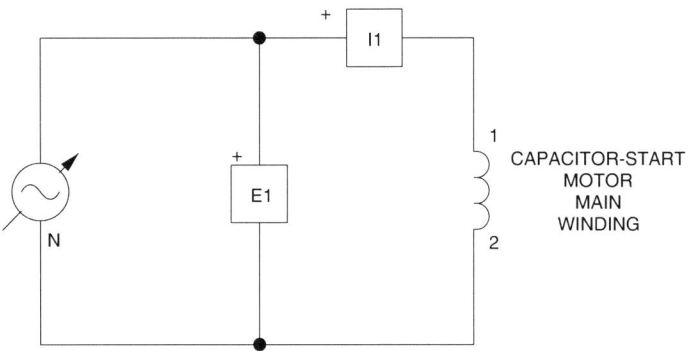

Figure 4-28. Capacitor-start motor circuit.

16. Turn the Power Supply on and set the voltage control knob to about 10%.

    In the Phasor Analyzer window, make the proper settings to observe the source voltage phasor (channel E1) and current phasor (channel I1). Observe that the current phasor (representing the main winding current) lags the source voltage phasor.

    On the Power Supply, set the voltage control knob to the 50% position.

    Does the capacitor-start motor start to rotate?

    ❏ Yes   ❏ No

17. Turn the Power Supply off and turn the voltage control knob fully counterclockwise.

    Connect the auxiliary winding of the Capacitor-Start Motor module as shown in Figure 4-29.

Ex. 4-4 – Single-Phase Induction Motors ◆ *Procedure*

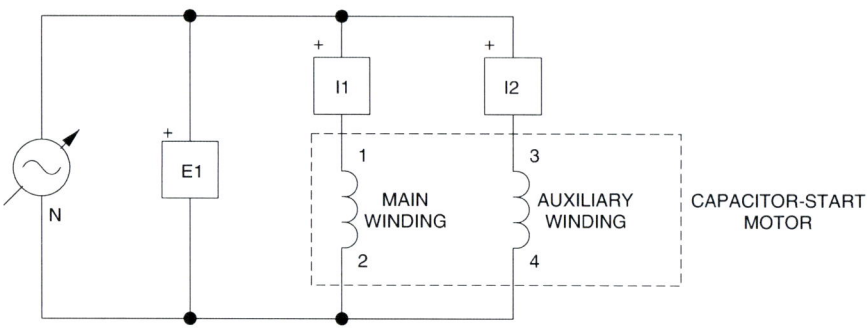

Figure 4-29. Connecting the auxiliary winding to the capacitor-start motor circuit.

18. Turn the Power Supply on and slowly set the voltage control knob to about 10%.

   Observe the current phasors (channels I1 and I2) in the Phasor Analyzer window. These phasors represent the main winding current and the auxiliary winding current, respectively.

   Is the phase shift of the auxiliary-winding current phasor with respect to the source voltage phasor (channel E1) less than that of the main-winding current phasor, thus confirming that the impedance of the auxiliary winding is more resistive and less inductive when the motor is starting?

   ❑ Yes   ❑ No

   Is the phase shift between the current phasors (channels I1 and I2) less than 90°?

   ❑ Yes   ❑ No

   On the Power Supply, set the voltage control knob to the 50% position.

   Does the capacitor-start motor start to rotate?

   ❑ Yes   ❑ No

    Since the nominal current of the auxiliary winding of the Capacitor-Start Motor may be exceeded while performing this manipulation, it is suggested to complete the manipulation within a time interval as short as possible.

   However, if the circuit breaker on the Capacitor-Start Motor trips, turn the Power Supply off, reset the breaker, turn the Power Supply on and continue the manipulation.

19. Turn the Power Supply off and turn the voltage control knob fully counterclockwise.

   Modify the capacitor-start motor circuit by connecting the capacitor on the Capacitor-Start Motor module in series with the auxiliary winding as shown in Figure 4-30.

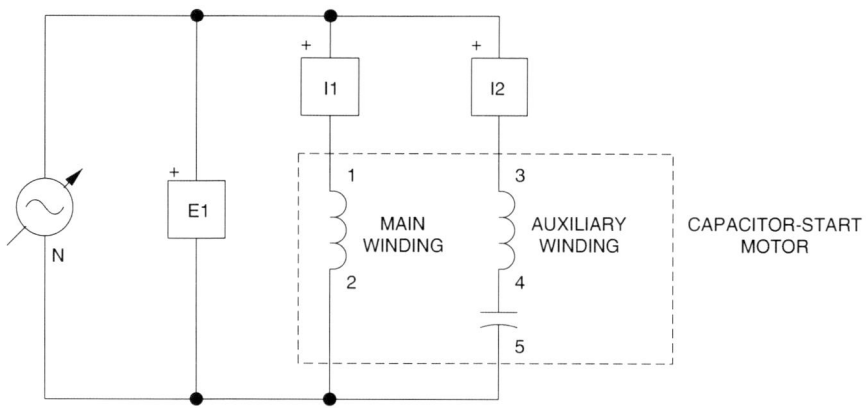

Figure 4-30. Connecting a capacitor in series with the auxiliary winding.

20. Turn the Power Supply on and slowly set the voltage control knob to about 10%.

Observe the current phasors (channels I1 and I2) in the Phasor Analyzer window.

Does connecting a capacitor in series with the auxiliary winding create a phase shift of approximately 90° between the current phasors?

❏ Yes   ❏ No

On the Power Supply, set the voltage control knob to the 50% position.

Does the capacitor-start motor start to rotate?

❏ Yes   ❏ No

Let the motor operate during a few minutes while observing the current phasors (on channels I1 and I2) in the Phasor Analyzer window.

Describe what happens.

_____

_____

_____

21. Turn the Power Supply off and turn the voltage control knob fully counterclockwise.

On the Capacitor-Start Motor, reset the tripped circuit breaker.

Modify the capacitor-start motor circuit by connecting the centrifugal switch on the Capacitor-Start Motor module in series with the auxiliary winding and the capacitor as shown in Figure 4-31.

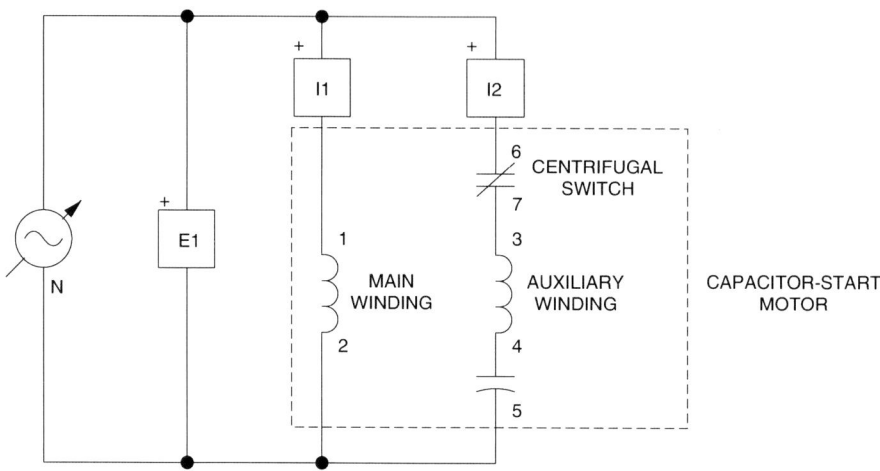

Figure 4-31. Connecting a centrifugal switch in series with the auxiliary winding and capacitor.

22. Turn the Power Supply on and slowly set the voltage control knob to 100%. While doing this, observe the current phasors (on channels I1 and I2) in the Phasor Analyzer window as the voltage increases.

    Does the capacitor-start motor start to rotate?

    ❏ Yes   ❏ No

    Briefly explain why the auxiliary-winding current phasor (channel I2) disappears a little after the motor has started to rotate.

    _____

    _____

23. On the Power Supply, turn the voltage control knob fully counterclockwise then set the 24 V - AC power switch to the O (off) position.

    Remove all leads and cables.

Ex. 4-4 – Single-Phase Induction Motors ♦ *Conclusion*

CONCLUSION

In this exercise, you observed that a three-phase squirrel-cage induction motor starts and runs almost normally when powered by only two phases of a three-phase ac power source, because a rotating magnetic field is maintained. However, you saw that when only one phase is connected to the motor, there is no rotating magnetic field and the motor is not able to start rotating. You demonstrated that adding an auxiliary winding and a capacitor to an induction motor allows it to start and run normally when powered by a single-phase ac power source. You saw that this produces two currents (the main- and auxiliary-winding currents) that are phase shifted of approximately 90°, and that these currents produce the necessary rotating magnetic field when the motor is starting. Finally, you observed that a centrifugal switch is used to disconnect the auxiliary winding when the single-phase induction motor reaches sufficient speed to maintain the rotating magnetic field.

REVIEW QUESTIONS

1. When only two phases are connected to a three-phase squirrel-cage induction motor, it

    a. runs almost normally.
    b. turns in the opposite direction.
    c. does not start.
    d. affects the amount of reactive power supplied by the motor.

2. When only one phase is connected to a three-phase squirrel-cage induction motor, it

    a. runs almost normally.
    b. turns in the opposite direction.
    c. does not start or it starts rotating abnormally.
    d. affects the amount of reactive power supplied by the motor.

3. An auxiliary winding and a capacitor are added to a single-phase induction motor to help

    a. it start.
    b. to increase the starting torque.
    c. to produce a phase shift between the winding currents.
    d. All of the above.

4. Single-phase induction motors of the capacitor-start type use a centrifugal switch to

    a. add an auxiliary winding and a capacitor to the motor circuit.
    b. remove an auxiliary winding and a capacitor from the motor circuit.
    c. add resistance only to the starting circuit.
    d. remove resistance only from the starting circuit.

5. The auxiliary winding has fewer turns of finer wire and therefore has

   a. lower resistance and higher inductance.
   b. lower resistance and lower inductance.
   c. higher resistance and higher inductance.
   d. higher resistance and lower inductance.

Unit Test

1. When a magnet moves above a conducting ladder, the currents induced in the ladder produces a magnetic field. This field interacts with the magnetic field of the magnet to produce a force that

   a. pulls the ladder in the direction opposite to the direction of the moving magnet.
   b. pulls the ladder in the same direction as the moving magnet.
   c. brakes the ladder.
   d. None of the above.

2. In a three-phase squirrel-cage induction motor, the rotating magnetic field is produced by

   a. two electromagnets placed at right angle from each other and two sine wave currents phase shifted of 90° from one another.
   b. three electromagnets placed at 90° from each other and three sine wave currents phase shifted of 120° from one another.
   c. three electromagnets placed at 120° from each other and three sine wave currents phase shifted of 120° from one another.
   d. three electromagnets placed at 120° from each other and three sine wave currents phase shifted of 90° from one another.

3. The slip is the difference between the

   a. synchronous speed and the rotor speed of an induction motor.
   b. no-load and full-load speeds of an induction motor.
   c. synchronous speed and the full-load speed of an induction motor.
   d. synchronous speed and the no-load speed of an induction motor.

4. The torque developed by an induction motor is zero when it rotates at synchronous speed because

   a. the full-load speed is equal to the synchronous speed.
   b. squirrel-cage reaction occurs.
   c. no currents are induced in the rotor.
   d. the currents induced in the rotor cancel out.

5. The direction of rotation of a three-phase squirrel-cage induction motor depends on the

   a. residual magnetism in the squirrel-cage rotor.
   b. interaction of the stator and rotor magnetic fields.
   c. design of the induction motor.
   d. phase sequence of the voltage applied to the motor stator windings.

6. When a conducting ladder moves below a fixed magnet, the currents induced in the ladder produces a magnetic field. This field interacts with the magnetic field of the magnet to produce a force that

   a. changes the direction in which the ladder moves.
   b. pulls the ladder in the same direction.
   c. brakes the ladder.
   d. None of the above.

7. A three-phase squirrel-cage induction motor connected to a three-phase power source

   a. cannot operate as an asynchronous generator.
   b. operates as a motor when it rotates at speeds higher than the synchronous speed.
   c. operates as an asynchronous generator when it rotates at speeds higher than the synchronous speed.
   d. always operates as an asynchronous generator because it always requires reactive power.

8. A three-phase squirrel-cage induction motor produces a torque of 30 N·m (266 lbf·in) at a speed of 1250 r/min when the voltage applied to each winding is 300 V. What torque will be developed at the same speed if the voltage is reduced to 250 V?

   a. 20.8 N·m (184 lbf·in)
   b. 25.0 N·m (221 lbf·in)
   c. 36.0 N·m (319 lbf·in)
   d. 43.0 N·m (382 lbf·in)

9. When a single-phase squirrel-cage induction motor rotates, the rotating magnetic field is produced by the interaction

   a. between the main and auxiliary stator windings
   b. of the magnetic fields in the main stator winding and the rotor.
   c. of the magnetic fields in the auxiliary stator winding and the rotor.
   d. between the capacitor, the main stator winding, and the rotor.

10. An auxiliary winding is required in a single-phase induction motor to

    a. reduce squirrel-cage reaction.
    b. allow the motor to produce starting torque.
    c. reduce the reactive power required to produce the rotating magnetic field.
    d. allow the motor to rotate at higher speeds.

Unit 5

Synchronous Motors

UNIT OBJECTIVE

After completing this unit, you will be able to demonstrate and explain the operating characteristics of synchronous motors using the Synchronous Motor/Generator module.

DISCUSSION OF FUNDAMENTALS

The principles of operation of the three-phase synchronous motor are very similar to those of the three-phase squirrel-cage induction motor. The stator is usually built in the same way (refer to Figure 4-4), and it creates a rotating magnetic field the same as illustrated in Figure 4-6. The rotor of the synchronous motor, however, is not a squirrel-cage construction, but rather a permanent magnet or an electromagnet installed on the motor shaft, as shown in Figure 5-1. This rotor is pulled along by the rotating magnetic field exactly as shown in Unit 1.

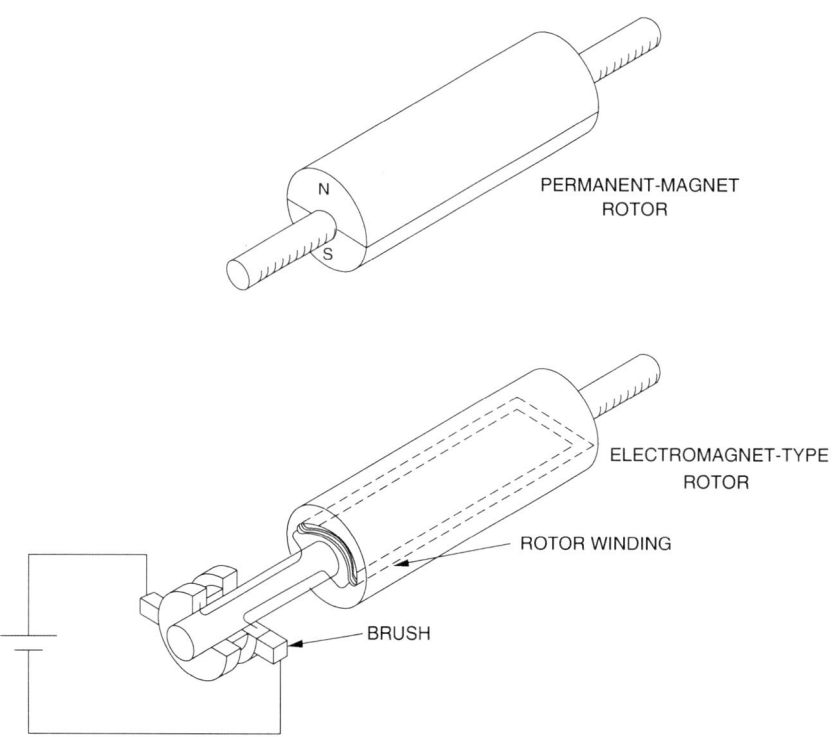

Figure 5-1. Rotor of a synchronous motor.

There is, however, a problem when starting a synchronous motor. When three-phase ac power is applied to the stator windings, a rotating magnetic field at synchronous speed n_S is immediately created. Since the rotor is at rest, it cannot catch up to the rotating magnetic field and the resulting torque acting on the rotor is fairly weak.

One way to aid in starting a synchronous motor having a rotor of the electromagnet type is to add a squirrel-cage to the rotor. During start-up, power is removed from the rotor electromagnet and three-phase ac power is applied to the stator windings. A rotating magnetic field is created, currents are induced in the squirrel cage, and the motor starts to rotate like a conventional three-phase squirrel-cage induction motor. When the motor speed stabilizes, dc power is applied to the electromagnet and the rotor locks to the rotating magnetic field and turns at exactly the synchronous speed n_S.

A synchronous motor with a permanent-magnet rotor cannot be started this way because the permanent magnet cannot be turned off. In this case, a variable-frequency ac source is used to supply power to the stator windings of the permanent-magnet synchronous motor. The frequency of the ac source is first set to a low value. This creates a stator magnetic field that rotates at a low speed, and thereby, allows the rotor to catch up to this field. The frequency of the ac source is then increased gradually to increase the speed to the desired value.

Exercise 5-1

The Three-Phase Synchronous Motor

EXERCISE OBJECTIVE

When you have completed this exercise, you will you will be able to demonstrate how to start a synchronous motor as well as some characteristics of a synchronous motor using the Synchronous Motor/Generator module.

DISCUSSION

The most interesting features of the three-phase synchronous motor are its ability to operate at exactly the same speed as the rotating magnetic field, the capability of running at unity power factor, and to be able to supply reactive power to an ac power source. As seen in Unit 4, an asynchronous motor always consumes reactive power, whether it operates as a motor or a generator. This is because the squirrel-cage induction motor requires reactive power to produce the rotating magnetic field. In the case of the three-phase synchronous motor, the rotating magnetic field is the sum of the magnetic fields produced by the stator and the rotor. If the rotor field is weak, the stator must contribute almost all the reactive power for the rotating magnetic field. The motor thus consumes reactive power like an inductor or an asynchronous motor. However, if the rotor field is strong, the stator acts to decrease the resulting field, and the motor thus supplies reactive power like a capacitor.

A graph of the reactive power Q versus the field current I_f (current in the rotor electromagnet) of a three-phase synchronous motor operating without load is shown in Figure 5-2. When the field current I_f is minimum, the magnetic field produced by the rotor is weak and the motor consumes a maximum of reactive power (Q is positive). The reactive power that is consumed decreases to zero as current I_f increases because the strength of the magnetic field produced by the rotor increases. When current I_f exceeds a certain value that depends on the characteristics of the motor, the rotor magnetic field is so strong that the motor starts to supply reactive power, i.e., Q becomes negative as illustrated in Figure 5-2.

The graph of the reactive power Q versus the field current I_f shows that a three-phase synchronous motor without load behaves like a three-phase reactive load whose nature (inductive or capacitive) and value depend on the field current I_f. Therefore, three-phase synchronous motors without load are also known as synchronous condensers when used to control the power factor on three-phase power networks.

Ex. 5-1 – The Three-Phase Synchronous Motor ♦ *Discussion*

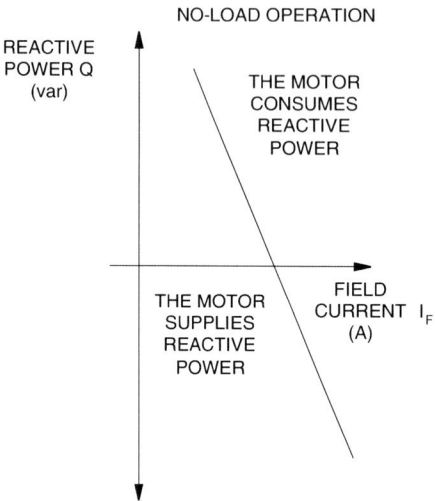

Figure 5-2. Reactive power Q versus field current I_f for a three-phase synchronous motor operating without load.

The graph of the line current I_L versus the field current I_f for a three-phase synchronous motor is a "V" type curve like that shown in Figure 5-3. This graph shows that the line current to the motor can be minimized by setting the field current I_f to the appropriate value. The field current required to minimize the line current is the same as that required to decrease the reactive power to zero. Therefore, the motor reactive power is zero when the line current is minimum.

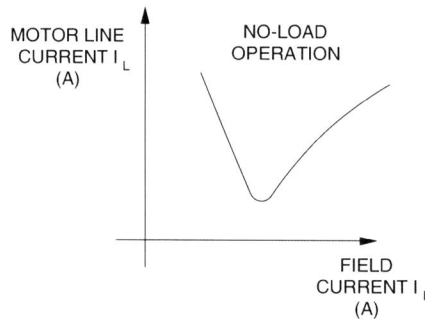

Figure 5-3. Line current I_L versus field current I_f for a three-phase synchronous motor operating without load.

The most inconvenient aspect of a three-phase synchronous motor is that it does not start easily, as is explained earlier in this unit.

Procedure summary

In the first part of the exercise, you will set up the equipment in the Workstation, connect the equipment as shown in Figure 5-4, and make the appropriate settings on the equipment.

In the second part of the exercise, you will see how to start a three-phase synchronous motor with a rotor of the electromagnet type. You will also vary the field current to see if this affects the motor speed and line current.

In the third part of the exercise, you will vary the field current by steps. For each step, you will record in the Data Table various electrical parameters related to the three-phase synchronous motor. You will then use this data to plot various graphs and determine many of the characteristics of the three-phase synchronous motor.

EQUIPMENT REQUIRED

Refer to the Equipment Utilization Chart in Appendix C to obtain the list of equipment required for this exercise.

PROCEDURE

 High voltages are present in this laboratory exercise. Do not make or modify any banana jack connections with the power on unless otherwise specified.

Setting up the equipment

1. Install the equipment required in the EMS workstation.

    Mechanically couple the prime mover/dynamometer module to the Synchronous Motor/Generator.

2. On the Power Supply, make sure the main power switch is set to the O (off) position, and the voltage control knob is turned fully counterclockwise. Ensure the Power Supply is connected to a three-phase power source.

     If you are using the Four-Quadrant Dynamometer/Power Supply, Model 8960-2, connect its POWER INPUT to a wall receptacle.

3. Ensure that the data acquisition module is connected to a USB port of the computer.

    Connect the POWER INPUT of the data acquisition module to the 24 V - AC output of the Power Supply.

     If you are using the Prime Mover/Dynamometer, Model 8960-1, connect its LOW POWER INPUT to the 24 V - AC output of the Power Supply.

    On the Power Supply, set the 24 V - AC power switch to the I (on) position.

     If you are using the Four-Quadrant Dynamometer/Power Supply, Model 8960-2, turn it on by setting its POWER INPUT switch to the I (on) position. Press and hold the FUNCTION button 3 seconds to have uncorrected torque values on the display of the Four-Quadrant Dynamometer/Power Supply. The indication "NC" appears next to the function name on the display to indicate that the torque values are uncorrected.

4. Start the Data Acquisition software (LVDAC or LVDAM). Open setup configuration file ACMOTOR1.DAI.

   In the Metering window, select layout 2. Make sure that the continuous refresh mode is selected.

5. Connect the equipment as shown in Figure 5-4. Connect the three resistor sections on the Resistive Load module in parallel to implement resistor R_1.

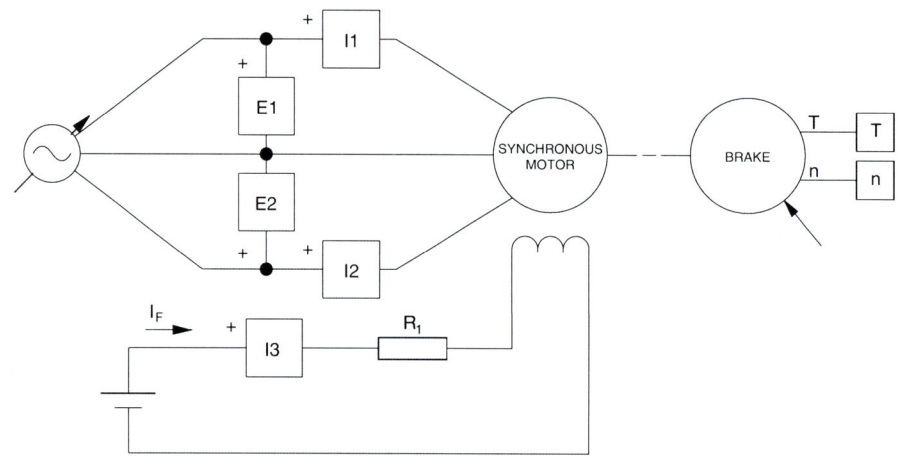

Local ac power network		R_1 (Ω)
Voltage (V)	Frequency (Hz)	
120	60	57
220	50	210
220	60	210
240	50	229

Figure 5-4. Synchronous motor coupled to a brake.

6. Set the Four-Quadrant Dynamometer/Power Supply or the Prime Mover/Dynamometer to operate as a brake, then set the brake torque control to maximum (fully clockwise position). To do this, refer to Exercise 1-1 or Exercise 1-2 if necessary.

    If you are performing the exercise using LVSIM®-EMS, you can zoom in the Prime Mover/Dynamometer module before setting the controls in order to see additional front panel markings related to these controls.

Starting a three-phase synchronous motor

7. In the Metering window, make sure that the torque correction function of the Torque meter is enabled.

    On the Synchronous Motor/Generator, set the EXCITER switch to the I (closed) position and turn the EXCITER knob fully clockwise.

    Turn the Power Supply on and set the voltage control knob so that the line voltage indicated by meter E line 1 is equal to the nominal line voltage of the synchronous motor.

     The rating of any of the machines is indicated in the lower left corner of the module front panel.

    Record the starting torque T_{START} of the synchronous motor (indicated by the Torque meter in the Metering window).

    T_{START} = _____ N · m (lbf · in)   (rotor electromagnet turned on)

8. On the Synchronous Motor/Generator, set the EXCITER switch to the O (open) position.

    Record the starting torque T_{START} of the synchronous motor in the following blank space.

    T_{START} = _____ N · m (lbf · in)   (rotor electromagnet turned off)

    Compare the starting torque obtained when the rotor electromagnet is turned off to that obtained when the rotor electromagnet is turned on.

    _____

    _____

    From the results obtained so far, would you conclude that it is desirable to turn the rotor electromagnet off before starting the synchronous motor? Briefly explain.

    _____

    _____

9. On the brake, slowly set the torque control to minimum (fully counterclockwise), wait until the synchronous motor speed stabilizes, and record the motor speed n (indicated by the Speed meter) in the following blank space.

    n = _____ r/min

On the Synchronous Motor/Generator, set the EXCITER knob to the mid position then set the EXCITER switch to the I (closed) position.

Does the motor speed n change?

❑ Yes   ❑ No

Record the motor speed n in the following blank space.

$n = $ _____ r/min

Is the motor speed now approximately equal to the nominal speed of the Synchronous Motor/Generator (synchronous speed n_S)?

❑ Yes   ❑ No

10. On the Synchronous Motor/Generator, slowly vary the setting of the EXCITER knob between the MIN. and MAX. positions to vary the field current I_f [indicated by meter I field (I_f)]. While doing this, observe the motor speed n and the motor line current I_{LINE} indicated by the Speed meter and meter I line 1, respectively.

Does varying the field current I_f vary the motor speed n?

❑ Yes   ❑ No

Does the motor line current I_{LINE} vary when the field current I_f is varied?

❑ Yes   ❑ No

On the Synchronous Motor/Generator, set the EXCITER knob to the MIN. position.

Characteristics of a three-phase synchronous motor

11. Change the value of resistor R_1 and vary the setting of the EXCITER knob on the Synchronous Motor/Generator so that the field current I_f [indicated by meter I field (I_f)] passes from the minimum current to the maximum current indicated in Table 5-1, in ten steps that are spaced as equally as possible. Note that it may be necessary to short circuit resistor R_1 to increase the field current to the maximum value indicated in the table. For each current setting, record the motor line voltage E_{LINE}, line current I_{LINE}, field current I_f, active power P, and reactive power Q in the Data Table. These parameters are indicated by meters E line 1, I line 1, I field (I_f), Act. power, and React. power, respectively.

Ex. 5-1 – The Three-Phase Synchronous Motor ♦ *Procedure*

Table 5-1. Range of field current.

Local ac power network		I_f (mA)
Voltage (V)	Frequency (Hz)	
120	60	300 to 900
220	50	100 to 500
220	60	100 to 500
240	50	100 to 500

12. When all data has been recorded, turn the voltage control knob fully counterclockwise, and turn the Power Supply off.

    In the Data Table window, confirm that the data has been stored, entitle the Data Table as DT511, and print the Data Table.

     Refer to the user guide dealing with the computer-based instruments for EMS to know how to edit, entitle, and print a Data Table.

13. In the Graph window, make the appropriate settings to obtain a graph of the motor line current I_{LINE} (obtained from meter I line 1) as a function of the field current I_f [obtained from meter I field (I_f)]. Entitle the graph as G511, name the x-axis as Synchronous motor field current, name the y-axis as Synchronous motor line current, and print the graph.

     Refer to the user guide dealing with the computer-based instruments for EMS to know how to use the Graph window of the Metering application to obtain a graph, entitle a graph, name the axes of a graph, and print a graph.

    Approximate the field current I_f that minimizes the motor line current I_{LINE} using graph G511. Record your result in the following blank space.

    $I_f =$ _____ A (for reducing the motor line current to minimum)

14. In the Graph window, make the appropriate settings to obtain a graph of the motor active power P and reactive power Q (obtained from meters Act. power and React. power, respectively) as a function of the field current I_f [obtained from meter I field (I_f)], using the data recorded previously in the Data Table. Entitle the graph as G511-1, name the x-axis as Synchronous motor field current, name the y-axis as Synchronous motor active power and reactive power, and print the graph.

    Does varying the field current I_f vary the active power consumed by the synchronous motor significantly?

    ❑ Yes   ❑ No

Ex. 5-1 – The Three-Phase Synchronous Motor ♦ *Procedure*

How does the motor reactive power Q vary when the field current I_f increases?

_____

_____

Could a synchronous motor operating without load be used to improve the power factor of a three-phase power network? Briefly explain.

_____

_____

15. Determine the field current I_f for which the reactive power Q is zero using graph G511-1. Record your result in the following blank space.

$I_f = $ _____ A (for reducing the motor reactive power to zero)

Compare the field current that sets the reactive power Q to zero with the field current that minimizes the motor line current I_{LINE}.

_____

_____

From the results obtained so far, can you conclude that the motor line current is minimum when the reactive power is zero?

❑ Yes ❑ No

16. On the Power Supply, set the 24 V - AC power switch to the O (off) position.

If you are using the Four-Quadrant Dynamometer/Power Supply, Model 8960-2, turn it off by setting its POWER INPUT switch to the O (off) position.

Remove all leads and cables.

Ex. 5-1 – The Three-Phase Synchronous Motor ♦ *Conclusion*

CONCLUSION

In this exercise, you saw that the rotor electromagnet must be turned off when starting a synchronous motor, to obtain a higher torque. You observed that once a synchronous motor rotates at a fairly high speed, the rotor electromagnet can be turned on to make the motor turn at the synchronous speed n_S. You found that varying the field current I_f of a synchronous motor (current in the rotor electromagnet) varies the motor line current I_{LINE} as well as the motor reactive power Q. You plotted graphs of the motor line current, active power P, and reactive power Q versus the field current. You found that the synchronous motor line current can be minimized by adjusting the field current. You observed that the synchronous motor can either sink or source reactive power depending on the value of the field current. You saw that this allows a three-phase synchronous motor to be used as a synchronous condenser to improve the power factor of a three-phase power network.

REVIEW QUESTIONS

1. The starting torque of a three-phase synchronous motor is increased when

    a. the rotor electromagnet is turned on.
    b. the rotor electromagnet is turned off.
    c. the power factor of the ac power network is unity.
    d. dc power is applied to one of the stator windings.

2. When a synchronous motor without load is connected to a three-phase ac power network, the resulting power factor depends on

    a. the speed of the motor.
    b. the active power consumed by the motor.
    c. the amount of field current.
    d. the line current.

3. Reactive power in a synchronous motor without load is minimum when the

    a. line current is maximum.
    b. line current is minimum.
    c. line current equals the field current.
    d. field current is minimum.

4. Synchronous condenser is another name for

    a. an asynchronous motor.
    b. a squirrel-cage motor.
    c. a split-phase motor.
    d. a synchronous motor operating without load.

5. The squirrel cage in a synchronous motor with a rotor of the electromagnet type

   a. minimizes the motor line current.
   b. prevents saturation of the rotor electromagnet.
   c. allows the motor to start when ac power is applied to the stator windings.
   d. makes the motor operate as a synchronous condenser.

Exercise 5-2

Synchronous Motor Pull-Out Torque

EXERCISE OBJECTIVE

When you have completed this exercise, you will be able to measure the pull-out torque of a synchronous motor using the Synchronous Motor/Generator and prime mover/dynamometer module.

DISCUSSION

One of the important characteristics of the three-phase synchronous motor shown in the previous exercise is that its speed is exactly the same as that of the stator rotating magnetic field (the synchronous speed n_S). When the synchronous motor operates without load torque, the electromagnet rotor is positioned so that its magnetic poles are aligned with those of the rotating magnetic field as shown in Figure 5-5a. However, when load torque is applied to the synchronous motor, the electromagnet rotor changes position with respect to the rotating magnetic field, i.e., the rotor falls behind the rotating magnetic field as shown in Figure 5-5b.

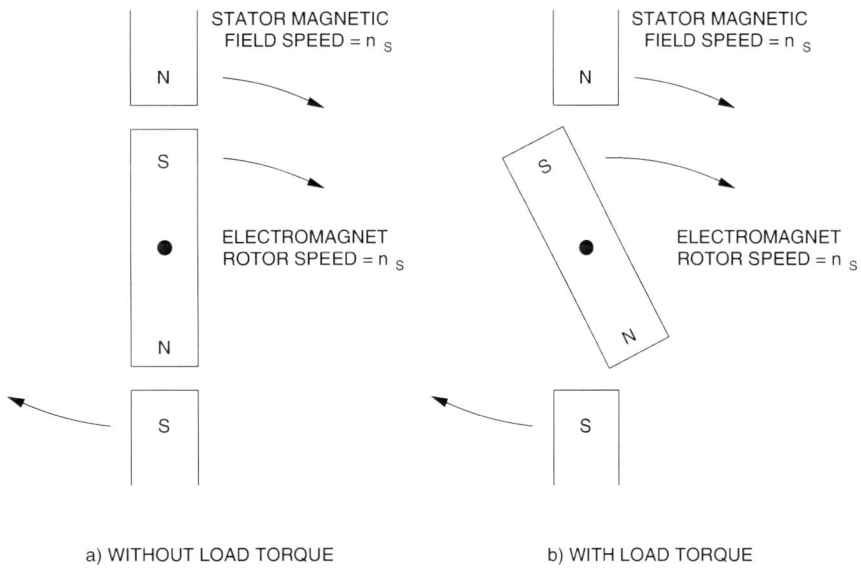

Figure 5-5. Effect of the load torque on the rotor position in a synchronous motor.

Ex. 5-2 – Synchronous Motor Pull-Out Torque ◆ *Procedure*

The lag of the rotor with respect to the rotating magnetic field of the stator is proportional to the load torque. Therefore, the higher the load torque, the further the rotor lags the rotating magnetic field. When the rotor lags the rotating magnetic field by 90°, it suddenly pulls out of synchronization with the rotating magnetic field, and the motor speed decreases greatly. Furthermore, the motor line current increases to high values and the motor vibrates. Protection devices should usually be installed on synchronous motors to ensure that the motor suffers no damage when synchronization is lost. The load torque at which synchronization is lost is called pull-out torque.

As might be imagined, higher values of field current I_f allow higher values of pull-out torque to be reached. The graph of pull-out torque versus field current I_f shown in Figure 5-6 indicates that the pull-out torque increases linearly as the field current I_f increases.

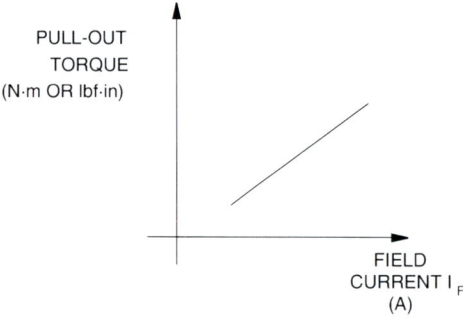

Figure 5-6. The pull-out torque increases linearly with the field current I_f.

Procedure summary

In the first part of the exercise, you will set up the equipment in the Workstation, connect the equipment as shown in Figure 5-7, and make the appropriate settings on the equipment.

In the second part of the exercise, you will set the field current I_f to various values and measure the pull-out torque. This will allow you to demonstrate how the field current I_f affects the pull-out torque.

EQUIPMENT REQUIRED

Refer to the Equipment Utilization Chart in Appendix C to obtain the list of equipment required for this exercise.

PROCEDURE

High voltages are present in this laboratory exercise. Do not make or modify any banana jack connections with the power on unless otherwise specified.

Ex. 5-2 – Synchronous Motor Pull-Out Torque ♦ *Procedure*

Setting up the equipment

1. Install the equipment in the EMS Workstation.

   Mechanically couple the prime mover/dynamometer module to the Synchronous Motor/Generator.

2. On the Power Supply, make sure the main power switch is set to the O (off) position, and the voltage control knob is turned fully counterclockwise. Ensure the Power Supply is connected to a three-phase power source.

    If you are using the Four-Quadrant Dynamometer/Power Supply, Model 8960-2, connect its POWER INPUT to a wall receptacle.

3. Ensure that the data acquisition module is connected to a USB port of the computer.

   Connect the POWER INPUT of the data acquisition module to the 24 V - AC output of the Power Supply.

    If you are using the Prime Mover/Dynamometer, Model 8960-1, connect its LOW POWER INPUT to the 24 V - AC output of the Power Supply.

   On the Power Supply, set the 24 V - AC power switch to the I (on) position.

    If you are using the Four-Quadrant Dynamometer/Power Supply, Model 8960-2, turn it on by setting its POWER INPUT switch to the I (on) position. Press and hold the FUNCTION button 3 seconds to have uncorrected torque values on the display of the Four-Quadrant Dynamometer/Power Supply. The indication "NC" appears next to the function name on the display to indicate that the torque values are uncorrected.

4. Start the Data Acquisition software (LVDAC or LVDAM). Open setup configuration file ACMOTOR1.DAI.

    If you are using LVSIM-EMS in LVVL, you must use the IMPORT option in the File menu to open the configuration file.

   In the Metering window, select layout 2. Make sure that the continuous refresh mode is selected.

5. Connect the equipment as shown in Figure 5-7.

   On the Synchronous Motor/Generator, make sure that the EXCITER switch is set to the O (open) position and the EXCITER knob is turned fully counterclockwise.

Ex. 5-2 – Synchronous Motor Pull-Out Torque ◆ Procedure

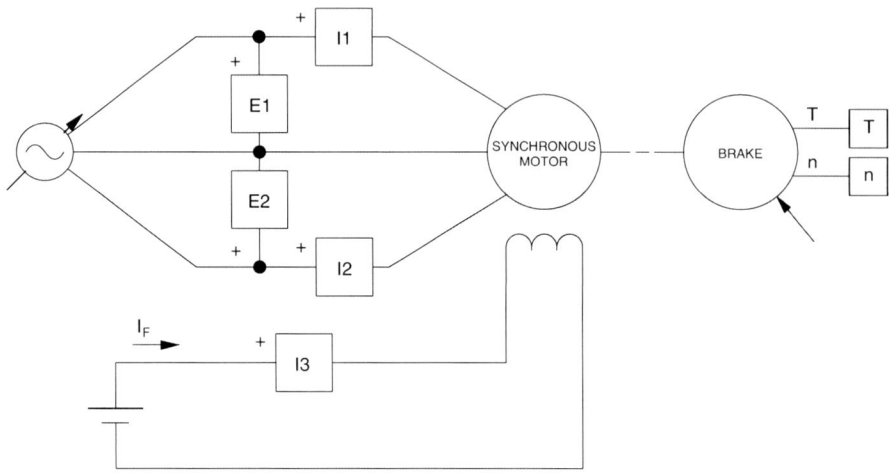

Figure 5-7. Synchronous motor coupled to a brake.

6. Set the Four-Quadrant Dynamometer/Power Supply or the Prime Mover/Dynamometer to operate as a brake, then set the brake torque control to minimum (fully counterclockwise position). To do this, refer to Exercise 1-1 or Exercise 1-2 if necessary.

 If you are performing the exercise using LVSIM®-EMS, you can zoom in on the Prime Mover/Dynamometer before setting the controls in order to see additional front panels markings related to these controls.

In the Metering window, make sure that the torque correction function of the Torque meter is enabled.

Synchronous motor pull-out torque

7. Turn the Power Supply on and set the voltage control knob so that the line voltage indicated by meter E line 1 is equal to the nominal line voltage of the synchronous motor. Wait for the speed of the synchronous motor to stabilize.

 The rating of any of the machines is indicated in the lower left corner of the module front panel.

On the Synchronous Motor/Generator, set the EXCITER switch to the I (closed) position. The synchronous motor should rotate at synchronous speed.

8. Record in the following blank space the field current I_f of the synchronous motor indicated by meter I field (I_f) in the Metering window.

$$I_f = \underline{} \text{ A}$$

Ex. 5-2 – Synchronous Motor Pull-Out Torque ◆ *Procedure*

On the brake, slowly increase the torque control until the synchronous motor pulls out of synchronization. While doing this, observe the synchronous motor torque and line current indicated by the Torque meter and meter I line 1, respectively. Record in the following blank spaces the motor torque when the motor pulls out of synchronization and the motor line current just before the motor pulls out of synchronization.

$T_{PULL\,OUT}$ = _____ N · m (lbf · in)

I_{LINE} = _____ A (motor in synchronization)

Record in the following blank spaces the synchronous motor line current and speed indicated by meters I line 1 and Speed, respectively.

I_{LINE} = _____ A (motor out of synchronization)

n = _____ r/min (motor out of synchronization)

Turn the Power Supply off.

On the Synchronous Motor/Generator, set the EXCITER switch to the O position.

On the brake, set the torque control to minimum (fully counterclockwise).

9. Describe how the speed varies when the synchronous motor pulls out of synchronization.

   _____
   _____

   How does the motor line current vary when the synchronous motor pulls out of synchronization?

   _____
   _____

Ex. 5-2 – Synchronous Motor Pull-Out Torque ♦ *Procedure*

10. Repeat steps 7 and 8 with the EXCITER knob on the Synchronous Motor/Generator set to the one-quarter, one-half, three-quarter, and maximum positions. For each setting of the EXCITER knob, record the values of the field current I_f and pull-out torque $T_{PULL\ OUT}$ in the following blank spaces.

    EXCITER knob set to one quarter of maximum:

    I_f = _____ A

    $T_{PULL\ OUT}$ = _____ N·m (lbf·in)

    EXCITER knob set to one half of maximum:

    I_f = _____ A

    $T_{PULL\ OUT}$ = _____ N·m (lbf·in)

    EXCITER knob set to three quarter of maximum:

    I_f = _____ A

    $T_{PULL\ OUT}$ = _____ N·m (lbf·in)

    EXCITER knob set to maximum:

    I_f = _____ A

    $T_{PULL\ OUT}$ = _____ N·m (lbf·in)

11. In the Data Table, insert five blank lines and then record the values of the field current I_f and pull-out torque $T_{PULL\ OUT}$ obtained in steps 8 and 10 in the columns Field current [I field (I_f)] and Pull-out torque (T), respectively.

    Entitle the Data Table as DT521, and print the Data Table.

     Refer to the user guide dealing with the computer-based instruments for EMS to know how to edit, entitle, and print a Data Table.

Ex. 5-2 – Synchronous Motor Pull-Out Torque ◆ *Conclusion*

12. In the Graph window, make the appropriate settings to obtain a graph of the pull-out torque $T_{PULL\ OUT}$ (obtained from the Torque meter) as a function of the field current I_f [obtained from meter I field (I_f)]. Entitle the graph as G521, name the x-axis as Synchronous motor field current, name the y-axis as Synchronous motor pull-out torque, and print the graph.

 Refer to the user guide dealing with the computer-based instruments for EMS to know how to use the Graph window of the Metering application to obtain a graph, entitle a graph, name the axes of a graph, and print a graph.

Does graph G521 demonstrate that the pull-out torque of the synchronous motor increases for higher values of field current?

❏ Yes   ❏ No

13. On the Power Supply, set the 24 V - AC power switch to the O (off) position.

 If you are using the Four-Quadrant Dynamometer/Power Supply, Model 8960-2, turn it off by setting its POWER INPUT switch to the O (off) position.

Remove all leads and cables.

CONCLUSION

In this exercise, you demonstrated the loss of synchronization between the rotor and the stator rotating magnetic field when the load on a synchronous motor is greater than the pull-out torque. You also observed that the pull-out torque is greater for higher values of field current.

REVIEW QUESTIONS

1. When load torque is applied to a synchronous motor

   a. the motor slows down.
   b. the motor speeds up.
   c. the rotor position falls behind the rotating magnetic field.
   d. the stator starts to rotate.

2. Pull-out torque is

   a. the minimum value of load torque that causes the nominal line current of a synchronous motor to be exceeded.
   b. the torque at which a synchronous motor pulls out of synchronization.
   c. the maximum torque for the minimum field current.
   d. the minimum torque that a synchronous motor can supply.

3. The synchronous motor in Figure 5-5 pulls out of synchronization when the rotor has shifted

   a. 30° behind the rotating magnetic field.
   b. 90° ahead of the rotating magnetic field.
   c. 30° ahead of the rotating magnetic field.
   d. 90° behind the rotating magnetic field.

4. When the field current in a synchronous motor is increased, the pull-out torque

   a. decreases.
   b. increases.
   c. does not change.
   d. increases momentarily until the speed stabilizes.

5. What happens when a synchronous motor loses synchronization?

   a. Nothing.
   b. The motor speeds up rapidly.
   c. The motor slows down, the line current increases, and the motor vibrates.
   d. The motor slows down and its torque increases.

Unit Test

1. A synchronous motor with a permanent-magnet rotor

   a. is started the same way as a synchronous motor with an electromagnet rotor.
   b. starts like a squirrel-cage induction motor.
   c. can be started using a variable-frequency ac power source.
   d. starts when dc power is applied to the rotor.

2. A three-phase synchronous motor draws reactive power from an ac power source. Decreasing the field current

   a. will increase the reactive power which the motor draws from the ac power source.
   b. will decrease the reactive power which the motor draws from the ac power source.
   c. will decrease the power factor of the motor.
   d. both a and c.

3. A three-phase synchronous motor supplies reactive power to an ac power source. Decreasing the field current

   a. will increase the reactive power which the motor supplies to the ac power source.
   b. will decrease the reactive power which the motor supplies to the ac power source.
   c. will decrease the power factor of the motor.
   d. both a and c.

4. A three-phase synchronous motor operates as a synchronous condenser. It is adjusted so that the power factor of the load connected to an ac power source is unity. One of the many inductive loads connected to the ac power source is removed. Therefore,

   a. the synchronous motor draws more reactive power from the ac power source.
   b. the synchronous motor supplies more reactive power to the ac power source.
   c. the field current of the synchronous motor should be decreased to readjust the power factor so that it is unity.
   d. the field current of the synchronous motor should be increased to readjust the power factor so that it is unity.

5. It is desirable to turn the rotor electromagnet of a synchronous motor off to

   a. obtain a higher starting torque.
   b. improve the power factor.
   c. increase the starting line current.
   d. increase the pull-out torque.

6. When the line current of a three-phase synchronous motor is minimized, the

   a. motor is used as a synchronous condenser.
   b. motor neither draws or supplies reactive power.
   c. field current is minimum.
   d. None of the above.

7. The pull-out torque of a synchronous motor depends on

   a. the power factor.
   b. the motor line current.
   c. the field current.
   d. None of the above.

8. The most interesting features of the three-phase synchronous motor are

   a. its ability to run at exactly the synchronous speed and to be able to operate as an asynchronous generator.
   b. its ability to run at exactly the synchronous speed and to be able to supply reactive power to an ac power source.
   c. the capability of running at unity power factor and to be able to draw reactive power from an ac power source.
   d. both b and c.

9. A three-phase synchronous motor operating without load acts as

   a. a resistive load whose value depends on the field current.
   b. an asynchronous generator operating without load.
   c. three independent single-phase power sources.
   d. a reactive load whose nature (inductive or capacitive) and value depend on the field current.

10. A three-phase synchronous motor

    a. can operate with either ac or dc power.
    b. does not start easily.
    c. is another type of ac induction motor.
    d. with a permanent-magnet rotor is often used as a synchronous condenser to adjust the power factor of an ac power source.

Unit 6

Three-Phase Synchronous Generators (Alternators)

UNIT OBJECTIVE

When you have completed this unit, you will to demonstrate and explain the operating characteristics of three-phase synchronous generators (alternators) using the Synchronous Motor/Generator and prime mover/dynamometer modules.

DISCUSSION OF FUNDAMENTALS

The three-phase synchronous generator, or alternator, produces most of the electricity used today. It is found in all electrical-power generating stations, whether they are of the hydroelectric, diesel, coal-fired, wind turbine, or nuclear type. The alternator also generates the electricity used in motor vehicles like cars and trucks.

The basic principle of operation for alternators is quite simple and can be explained using the simplified single-phase alternator shown in Figure 6-1. An electromagnet creates a magnetic field in the rotor. The electromagnet rotor is coupled to a source of mechanical power, such as a water turbine, to make it rotate. As a result, a continually-changing magnetic flux links the stator winding and induces an alternating voltage across the stator winding as shown in Figure 6-1.

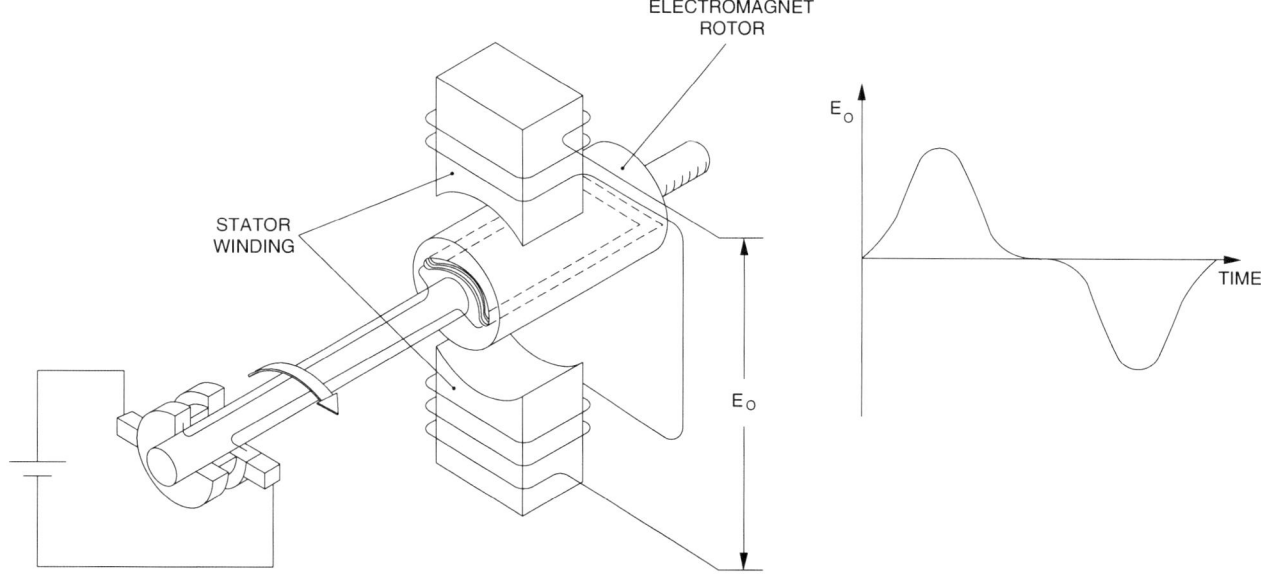

Figure 6-1. An alternating voltage is produced by the continually-changing magnetic flux linking the stator winding.

The way the conductors are wound in the stator of a synchronous generator determines the waveform of the voltage induced across the stator winding. The stator-winding conductors in synchronous generators are usually wound in such a way that the induced voltage has a sinusoidal waveform.

The stator in a three-phase synchronous generator is provided with three windings located at 120° from one another. As a result, three sine-wave voltages phase shifted by 120° with respect to each other are induced in the three stator windings. The stator of a three-phase synchronous generator is in fact very similar to the stator of a three-phase squirrel-cage induction motor shown in Figure 4-4.

Exercise 6-1

Synchronous Generator No-Load Operation

EXERCISE OBJECTIVE When you have completed this exercise, you will be able to demonstrate the no-load operation of a three-phase synchronous generator using the Synchronous Motor/Generator module.

DISCUSSION In three-phase synchronous generators, the stronger the rotor electromagnet, the greater the magnetic flux linking the stator windings, and the higher the alternating voltages induced across the stator windings. Furthermore, since the induced voltages are proportional to the rate of change of the magnetic flux linking the stator windings, one can easily deduce that the faster the rotor turns, the higher the amplitude of the induced voltages. In brief, the amplitude of the voltages produced by a three-phase synchronous generator is proportional to the strength of the rotor electromagnet and the rotation speed.

There is a direct relationship between the speed of the rotor and the frequency of the voltage induced across each stator winding of a synchronous generator. When the rotor of the synchronous generator shown in Figure 6-1 rotates at a speed of one revolution per second, the frequency of the induced voltage is one hertz. Since speed is usually expressed in revolutions per minute, the equation relating the speed of rotation to the frequency of the voltage produced by the synchronous generator shown in Figure 6-1 is as follows.

$$f = \frac{n}{60} \quad \text{(for generators with a stator having a single pair of poles)}$$

where f is the frequency, expressed in hertz (Hz)
        n is the speed, expressed in revolutions per minute (r/min)

However, each stator winding in large synchronous generators usually has several north and south poles instead of just a single pair as illustrated in Figure 6-1. As a result, a higher frequency is obtained for a given speed of rotation. The frequency of synchronous generators, regardless of the number of pairs of north and south poles, is determined by simply multiplying the speed n in the previous equation by P, which is the number of pairs of poles of each stator winding. The equation for determining the frequency of the voltage produced by a synchronous generator is thus,

$$f = \frac{n \times P}{60} \quad \text{(for any types of synchronous generators)}$$

Ex. 6-1 – Synchronous Generator No-Load Operation ♦ *Procedure*

Note that the Synchronous Motor/Generator has two north poles and two south poles per stator winding, thus two pairs of poles per stator winding. Therefore, P equals 2 for the Synchronous Motor/Generator.

Although small technical differences exist between a synchronous machine designed to operate as a motor and a synchronous machine designed to operate as a generator, both modes of operation can be demonstrated using a same synchronous machine.

Procedure summary

In the first part of the exercise, you will set up the equipment in the Workstation, connect the equipment as shown in Figure 6-2, and make the appropriate settings on the equipment.

In the second part of the exercise, you will vary the speed and field current and observe how this affects the no-load operation of a three-phase synchronous generator.

In the third part of the exercise, you will vary the field current of the synchronous generator by steps. For each step, you will record in the Data Table various electrical parameters related to the three-phase synchronous generator. You will also vary the speed of the synchronous generator by steps while recording various electrical parameters related to the synchronous generator. You will use the recorded data to plot various graphs and determine many of the characteristics of the three-phase synchronous generator.

EQUIPMENT REQUIRED

Refer to the Equipment Utilization Chart in Appendix C to obtain the list of equipment required for this exercise.

PROCEDURE

 High voltages are present in this laboratory exercise. Do not make or modify any banana jack connections with the power on unless otherwise specified.

Setting up the equipment

1. Install the equipment required in the EMS workstation.

   Mechanically couple the prime mover/dynamometer module to the Synchronous Motor/Generator.

2. On the Power Supply, make sure the main power switch is set to the O (off) position, and the voltage control knob is turned fully counterclockwise. Ensure the Power Supply is connected to a three-phase power source.

    If you are using the Four-Quadrant Dynamometer/Power Supply, Model 8960-2, connect its POWER INPUT to a wall receptacle.

Ex. 6-1 – Synchronous Generator No-Load Operation ◆ *Procedure*

3. Ensure that the data acquisition module is connected to a USB port of the computer.

   Connect the POWER INPUT of the data acquisition module to the 24 V - AC output of the Power Supply.

    If you are using the Prime Mover/Dynamometer, Model 8960-1, connect its LOW POWER INPUT to the 24 V - AC output of the Power Supply.

   On the Power Supply, set the 24 V - AC power switch to the I (on) position

    If you are using the Four-Quadrant Dynamometer/Power Supply, Model 8960-2, turn it on by setting its POWER INPUT switch to the I (on) position. Press and hold the FUNCTION button 3 seconds to have uncorrected torque values on the display of the Four-Quadrant Dynamometer/Power Supply. The indication "NC" appears next to the function name on the display to indicate that the torque values are uncorrected.

4. Start the Data Acquisition software (LVDAC or LVDAM). Open setup configuration file ACMOTOR1.DAI.

   In the Metering window, select layout 2. Make sure that the continuous refresh mode is selected.

5. Connect the equipment as shown in Figure 6-2.

   On the Synchronous Motor/Generator, set the EXCITER switch to the I (closed) position and the EXCITER knob to three quarters of maximum.

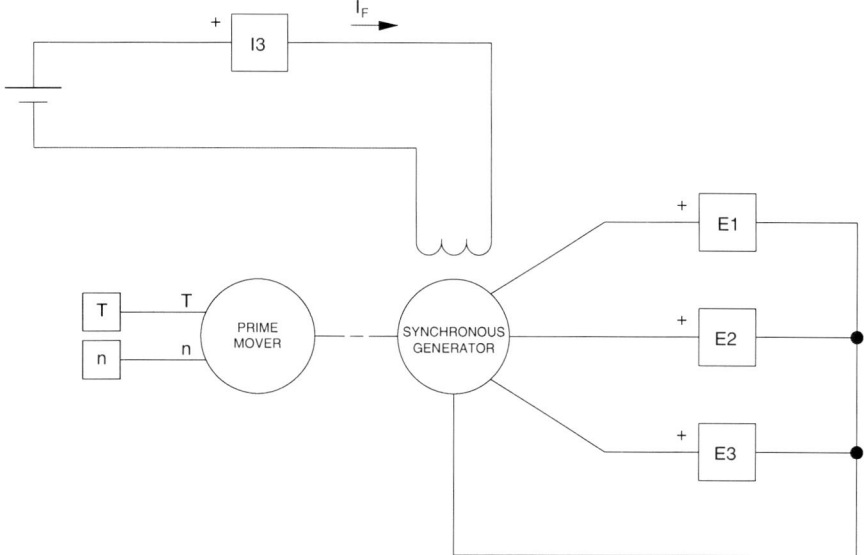

Figure 6-2. Synchronous generator coupled to a prime mover.

Ex. 6-1 – Synchronous Generator No-Load Operation ♦ *Procedure*

6. Set the Four-Quadrant Dynamometer/Power Supply or the Prime Mover/Dynamometer to operate as a clockwise prime mover. To do this, refer to Exercise 1-1 or Exercise 1-2 if necessary.

If you are performing the exercise using LVSIM®-EMS, you can zoom in the Prime Mover/Dynamometer module before setting the controls in order to see additional front panel markings related to these controls.

Synchronous generator no-load operation

7. Turn the Power Supply on. Set the prime mover speed (indicated by the Speed meter in the Metering window) so that it is equal to the nominal speed of the Synchronous Motor/Generator.

The rating of any of the machines is indicated in the lower left corner of the module front panel.

In the Oscilloscope window, make the appropriate settings to observe the waveforms of the voltages on channels E1, E2, and E3, that is, the voltages induced across each of the stator windings of the synchronous generator.

Are the waveforms sinusoidal?

❏ Yes   ❏ No

What is the approximate phase shift φ between each of the voltage waveforms?

$\varphi = $ _____ °

8. In the Oscilloscope window, select the display continuous-refresh function.

Slowly adjust the prime mover speed until it is approximately equal to 1000 r/min. While doing this, observe the waveforms of the voltages on channels E1, E2, and E3 in the Oscilloscope window.

How do the amplitude and frequency of the voltage waveforms vary when the speed of the synchronous generator is decreased? Briefly explain why.

_____
_____
_____

Does varying the speed of the synchronous generator affect the phase shift between the voltage waveforms? Why?

_____
_____

Ex. 6-1 – Synchronous Generator No-Load Operation ◆ *Procedure*

9. On the Synchronous Motor/Generator, slowly turn the EXCITER knob counterclockwise to decrease the field current I_f. While doing this, observe the waveforms of voltages on channels E1, E2, and E3 of the Oscilloscope.

   How does the amplitude of the voltage waveforms vary when the field current I_f of the synchronous generator is decreased? Briefly explain why.

   _____

   _____

   _____

   Does varying the field current I_f of the synchronous generator affect the frequency of the voltage waveforms and the phase shift between the voltage waveforms? Why?

   _____

   _____

   Set the prime mover speed to 0, then turn the Power Supply off.

Ex. 6-1 – Synchronous Generator No-Load Operation ◆ *Procedure*

Characteristics of a synchronous generator

10. Modify the connections so that the modules are connected as shown in Figure 6-3. Connect the three resistor sections on the Resistive Load module in parallel to implement resistor R_1.

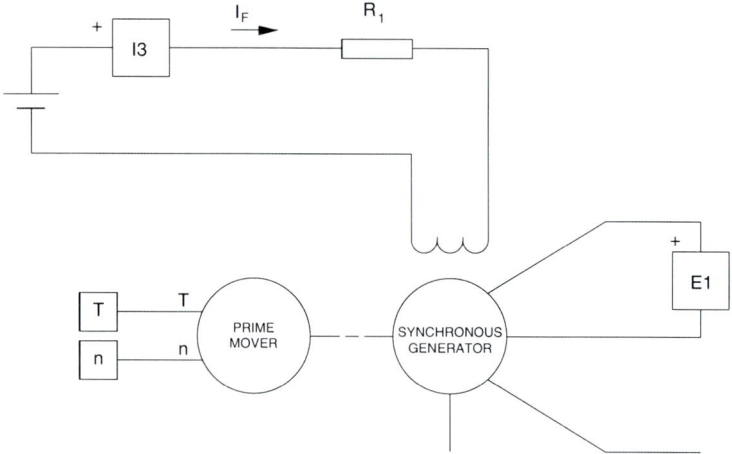

Local ac power network		R_1 (Ω)
Voltage (V)	Frequency (Hz)	
120	60	∞
220	50	∞
220	60	∞
240	50	∞

Figure 6-3. Circuit used to observe the no-load operation of a synchronous generator.

11. The Frequency meter in the Metering window will be used to measure the frequency of the voltage produced by the synchronous generator.

    Turn the Power Supply on and set the prime mover speed so that it is equal to the nominal speed of the Synchronous Motor/Generator.

12. Record the synchronous generator output voltage E_O, field current I_f, frequency f, and speed n in the Data Table. These parameters are indicated by meters E line 1, I field (I_f), Frequency, and Speed, respectively.

    Change the value of resistor R_1 and vary the setting of the EXCITER knob on the Synchronous Motor/Generator to increase the field current I_f to the value indicated in Table 6-1, in 10 steps that are spaced as equally as possible. Note that it may be necessary to short circuit resistor R_1 to increase the field current to the maximum value indicated in Table 6-1. For each current setting, readjust the prime mover speed so that it remains equal to the nominal speed of the Synchronous Motor/Generator, then record the data in the Data Table.

Table 6-1. Field current.

Local ac power network		I_f (mA)
Voltage (V)	Frequency (Hz)	
120	60	750
220	50	450
220	60	450
240	50	450

13. Short circuit resistor R_1 using a connection lead.

    On the Synchronous Motor/Generator, turn the EXCITER knob fully clockwise to set the field current I_f to maximum.

    Readjust the prime mover speed so that it remains equal to the nominal speed of the Synchronous Motor/Generator.

    Record the data in the Data Table.

    Set the prime mover speed to 0, then turn the Power Supply off.

14. In the Data Table window, confirm that the data has been stored, entitle the Data Table as DT611, and print the Data Table.

     Refer to the user guide dealing with the computer-based instruments for EMS to know how to edit, entitle, and print a Data Table.

15. Record the frequency of the voltages produced by the synchronous generator in the following blank space. This frequency is indicated in the column "Frequency" of the Data Table.

    $f = $ _____ Hz (measured)

    Calculate the theoretical frequency of the voltages produced by the Synchronous Motor/Generator using the following equation. (P is the number of pairs of poles per stator winding, i.e. 2).

    $$f = \frac{n \times P}{60} = \underline{\hspace{1cm}} = \underline{\hspace{1cm}} \text{ Hz}$$

    Compare the measured and calculated frequencies. Are they approximately equal?

    ❑ Yes   ❑ No

Ex. 6-1 – Synchronous Generator No-Load Operation ◆ Procedure

16. In the Graph window, make the appropriate settings to obtain a graph of the synchronous generator output voltage E_O (obtained from meter E line 1) as a function of the field current I_f [obtained from meter I field (I_f)]. Entitle the graph as G611, name the x-axis as Synchronous generator field current, name the y-axis as Synchronous generator output voltage, and print the graph.

 Refer to the user guide dealing with the computer-based instruments for EMS to know how to use the Graph window of the Metering application to obtain a graph, entitle a graph, name the axes of a graph, and print a graph.

Briefly explain why the relationship between the synchronous generator output voltage E_O and field current I_f is non linear for high values of current I_f.

___

___

___

In the Data Table window, clear the recorded data.

17. Turn the Power Supply on.

On the Synchronous Motor/Generator, set the EXCITER knob so that the field current I_f indicated by meter I field (I_f) is equal to the value given in Table 6-2.

Table 6-2. Field current of the synchronous generator.

Local ac power network		I_f (mA)
Voltage (V)	Frequency (Hz)	
120	60	500
220	50	300
220	60	300
240	50	300

18. Record the synchronous generator output voltage E_O, field current I_f, frequency f, and speed n in the Data Table. These parameters are indicated by meters E line 1, I field (I_f), Frequency, and Speed, respectively.

Increase the prime mover speed to the value given in the following table by increments of 200 r/min. For each speed setting, record the data in the Data Table.

Table 6-3. Maximum speed.

Local ac power network		Maximum speed ($n_{MAX.}$)
Voltage (V)	Frequency (Hz)	
120	60	2400
220	50	2000
220	60	2000
240	50	1800

19. When all data has been recorded, set the prime mover speed to 0, then turn the Power Supply off.

In the Data Table window, confirm that the data has been stored, entitle the Data Table as DT612, and print the Data Table.

20. In the Graph window, make the appropriate settings to obtain a graph of the synchronous generator output voltage E_O (obtained from meter E line 1) as a function of the speed n (obtained from the Speed meter). Entitle the graph as G612, name the x-axis as Synchronous generator speed, name the y-axis as Synchronous generator output voltage, and print the graph.

Describe how the synchronous generator output voltage varies as the speed varies.

_____

_____

21. In the Graph window, make the appropriate settings to obtain a graph of the synchronous generator frequency f (obtained from the Frequency meter) as a function of the speed n (obtained from the Speed meter). Entitle the graph as G612-1, name the x-axis as Synchronous generator speed, name the y-axis as Synchronous generator frequency, and print the graph.

Describe how the frequency of the voltages produced by the synchronous generator varies as the speed varies.

_____

_____

22. On the Power Supply, set the 24 V - AC power switch to the O (off) position.

 If you are using the Four-Quadrant Dynamometer/Power Supply, Model 8960-2, turn it off by setting its POWER INPUT switch to the O (off) position.

Remove all leads and cables.

Ex. 6-1 – Synchronous Generator No-Load Operation ♦ *Conclusion*

CONCLUSION

In this exercise, you observed that a three-phase synchronous generator produces three sine-wave voltages that are phase shifted by 120° from each other. You saw that decreasing the synchronous generator speed decreases the amplitude and frequency of the sine-wave voltages. You observed that decreasing the field current of the synchronous generator decreases the amplitude of the sine-wave voltages. You plotted a graph of the synchronous generator output voltage versus the field current. You plotted graphs of the synchronous generator output voltage and frequency versus speed. These graphs showed that the output voltage and frequency are proportional to the synchronous generator speed.

REVIEW QUESTIONS

1. Most electrical power that is consumed today is produced by

   a. synchronous condensers.
   b. synchronous generators.
   c. alternators.
   d. both b and c.

2. When the speed of a synchronous generator is increased

   a. the output voltage increases and the frequency decreases.
   b. the output voltage decreases and the frequency increases.
   c. both the output voltage and frequency decrease.
   d. both the output voltage and frequency increase.

3. How does the field current affect the frequency of the voltages produced by a three-phase synchronous generator

   a. Frequency increases as I_f increases.
   b. Frequency decreases as I_f decreases.
   c. Frequency is not affected by changes in field current.
   d. both a and b.

4. Multiplying the speed of an alternator by $P/60$ allows the

   a. theoretical frequency to be determined.
   b. theoretical output voltage to be determined.
   c. theoretical field current to be determined.
   d. number of poles to be determined.

5. Alternator is another name for a

   a. three-phase synchronous motor.
   b. three-phase synchronous generator.
   c. three-phase synchronous condenser.
   d. three-phase ac-to-dc converter.

Exercise 6-2

Voltage Regulation Characteristics

EXERCISE OBJECTIVE

When you have completed this exercise, you will be able to demonstrate the voltage regulation characteristics of a synchronous generator using the Synchronous Motor/Generator module.

DISCUSSION

As was seen in Unit 2 of this manual, a dc generator can be represented by the simplified equivalent circuit shown in Figure 6-4. In this circuit, the voltage E_{EMF} depends on the speed at which the generator rotates and the strength of the field electromagnet. Resistor R_A represents the resistance of the armature conductors.

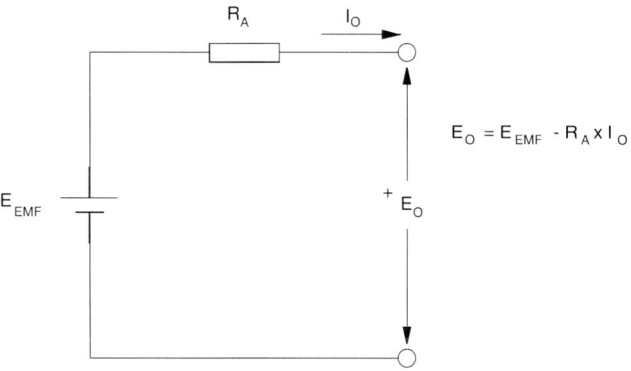

Figure 6-4. Simplified equivalent circuit of a dc generator.

A simplified equivalent circuit similar to that of the dc generator can be used to represent each phase of a three-phase synchronous generator. Figure 6-5 shows the simplified equivalent circuit for one phase of a three-phase synchronous generator. To represent a complete three-phase synchronous generator, three circuits like the one shown in Figure 6-5 would be used.

Ex. 6-2 – Voltage Regulation Characteristics ◆ *Discussion*

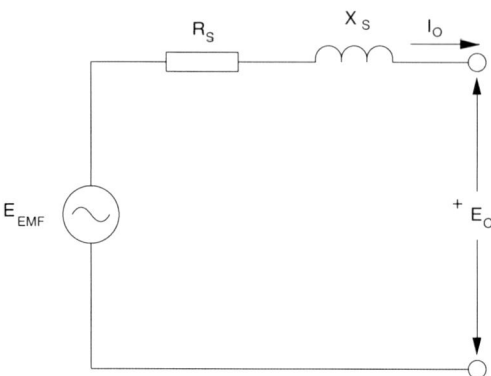

Figure 6-5. Simplified equivalent circuit for one phase of a three-phase synchronous generator.

As for a dc generator, the voltage E_{EMF} in the simplified circuit of the synchronous generator depends on the rotation speed as well as the strength of the electromagnet. Furthermore, there is a resistor (R_S) in the simplified circuit of the synchronous generator, as in the simplified circuit of the dc generator, that represents the resistance of the stator coil conductors. There is also an additional element in the simplified circuit of the synchronous generator, reactance X_S, which represents the inductive reactance of the stator coil conductors. Reactance X_S is known as the synchronous reactance of the synchronous generator and its value, expressed in ohms, is usually much greater than that of resistor R_S.

When the synchronous generator is operated at constant speed and with a fixed current in the rotor electromagnet (field current I_f), voltage E_{EMF} is constant and the equivalent circuit for each phase is very similar to that of a single-phase transformer, shown in Unit 7 of the student manual entitled *Power Circuits and Transformers*. Figure 6-6 shows voltage regulation characteristics (curves of the output voltage E_O versus the output current I_O) of a synchronous generator for resistive, inductive, and capacitive loads. These characteristics are very similar to those obtained with a single-phase transformer.

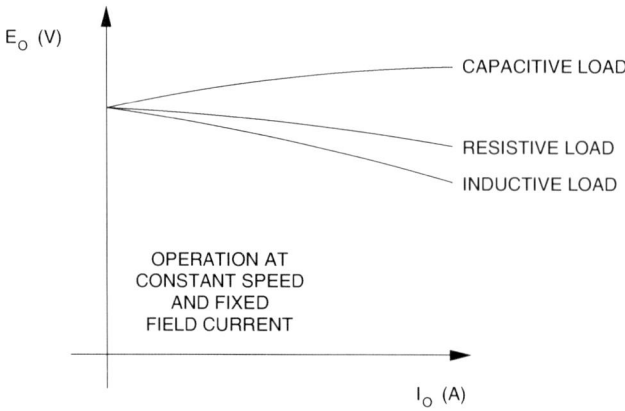

Figure 6-6. Voltage regulation characteristics of a synchronous generator.

Procedure summary

In the first part of the exercise, you will set up the equipment in the Workstation, connect the equipment as shown in Figure 6-7, and make the appropriate settings on the equipment.

In the second part of the exercise, you will set the speed of rotation and the field current of the synchronous generator. You will vary the value of the resistive load connected to the generator by steps while maintaining a constant speed. For each load value, you will record the synchronous generator output voltage, output current, field current, and speed. You will use the recorded data to plot a graph of the output voltage versus output current. You will then repeat this part of the exercise twice using an inductive load and a capacitive load.

EQUIPMENT REQUIRED

Refer to the Equipment Utilization Chart in Appendix C to obtain the list of equipment required for this exercise.

PROCEDURE

 High voltages are present in this laboratory exercise. Do not make or modify any banana jack connections with the power on unless otherwise specified.

Setting up the equipment

1. Install the equipment required in the EMS workstation.

   Mechanically couple the prime mover/dynamometer module to the Synchronous Motor/Generator.

2. On the Power Supply, make sure the main power switch is set to the O (off) position, and the voltage control knob is turned fully counterclockwise. Ensure the Power Supply is connected to a three-phase power source.

    If you are using the Four-Quadrant Dynamometer/Power Supply, Model 8960-2, connect its POWER INPUT to a wall receptacle.

3. Ensure that the data acquisition module is connected to a USB port of the computer.

   Connect the POWER INPUT of the data acquisition module to the 24 V - AC output of the Power Supply.

    If you are using the Prime Mover/Dynamometer, Model 8960-1, connect its LOW POWER INPUT to the 24 V - AC output of the Power Supply.

Ex. 6-2 – Voltage Regulation Characteristics ♦ *Procedure*

On the Power Supply, set the 24 V - AC power switch to the I (on) position

 If you are using the Four-Quadrant Dynamometer/Power Supply, Model 8960-2, turn it on by setting its POWER INPUT switch to the I (on) position. Press and hold the FUNCTION button 3 seconds to have uncorrected torque values on the display of the Four-Quadrant Dynamometer/Power Supply. The indication "NC" appears next to the function name on the display to indicate that the torque values are uncorrected.

4. Start the Data Acquisition software (LVDAC or LVDAM). Open setup configuration file ACMOTOR1.DAI.

    In the Metering window, select layout 2. Make sure that the continuous refresh mode is selected.

5. Connect the equipment as shown in Figure 6-7.

    On the Synchronous Motor/Generator, set the EXCITER switch to the I (closed) position and the EXCITER knob to the mid position.

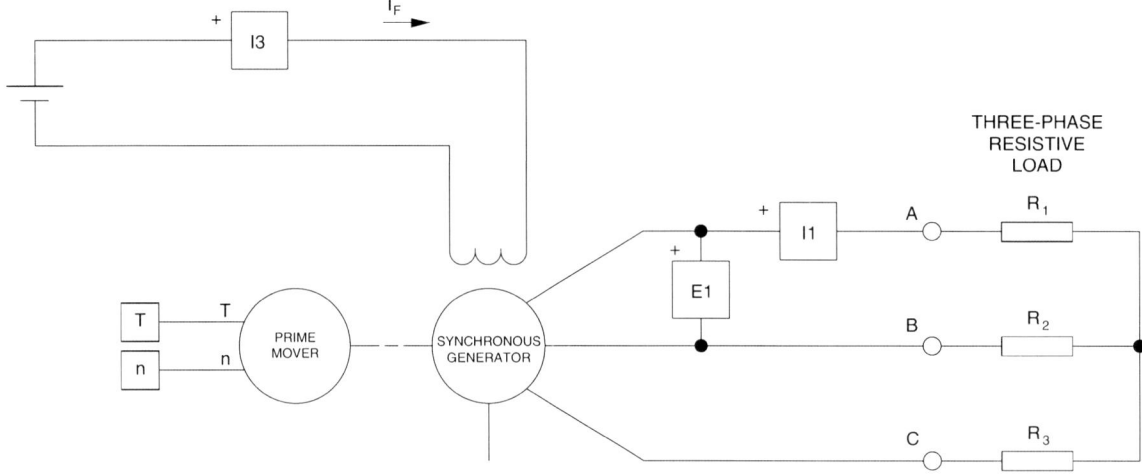

Local ac power network		R_1, R_2, R_3 (Ω)
Voltage (V)	Frequency (Hz)	
120	60	∞
220	50	∞
220	60	∞
240	50	∞

Figure 6-7. Synchronous generator under load coupled to a prime mover.

6. Set the Four-Quadrant Dynamometer/Power Supply or the Prime Mover/Dynamometer to operate as a clockwise prime mover. To do this, refer to Exercise 1-1 or Exercise 1-2 if necessary.

If you are performing the exercise using LVSIM®-EMS, you can zoom in the Prime Mover/Dynamometer module before setting the controls in order to see additional front panel markings related to these controls.

Voltage regulation characteristics

7. Turn the Power Supply on. Set the prime mover speed (indicated by the Speed meter in the Metering window) so that it is equal to the nominal speed of the Synchronous Motor/Generator.

The rating of any of the machines is indicated in the lower left corner of the module front panel.

8. On the Synchronous Motor/Generator, set the EXCITER knob so that the line-to-line output voltage E_O of the synchronous generator (indicated by meter E line 1 in the Metering window) is equal to the nominal value.

Record the synchronous generator output voltage E_O, output current I_O, field current I_f, and speed n in the Data Table. These parameters are indicated by meters E line 1, I line 1, I field (I_f), and Speed, respectively.

9. Modify the settings on the Resistive Load module so that the resistance of resistors R_1, R_2, and R_3 decreases by steps as indicated in Table 6-4. You can refer to Appendix B of this manual to know how to obtain the various resistance values given in Table 6-4. For each resistance setting, readjust the prime mover speed so that it remains equal to the nominal speed of the Synchronous Motor/Generator, then record the data in the Data Table.

Table 6-4. Decreasing the resistance of R_1, R_2, and R_3 to load the synchronous generator.

Local ac power network		R_1, R_2, R_3 (Ω)	R_1, R_2, R_3 (Ω)	R_1, R_2, R_3 (Ω)	R_1, R_2, R_3 (Ω)	R_1, R_2, R_3 (Ω)	R_1, R_2, R_3 (Ω)	R_1, R_2, R_3 (Ω)
Voltage (V)	Frequency (Hz)							
120	60	1200	600	400	300	240	200	171
220	50	4400	2200	1467	1100	880	733	629
220	60	4400	2200	1467	1100	880	733	629
240	50	4800	2400	1600	1200	960	800	686

10. When all data has been recorded, set the prime mover speed to 0, then turn the Power Supply off.

11. In the Data Table window, confirm that the data has been stored, entitle the Data Table as DT621, and print the Data Table.

Refer to the user guide dealing with the computer-based instruments for EMS to know how to edit, entitle, and print a Data Table.

12. In the Graph window, make the appropriate settings to obtain a graph of the synchronous generator output voltage E_O (obtained from meter E line 1) as a function of the output current I_O (obtained from meter I line 1). Entitle the graph as G621, name the x-axis as Synchronous generator output current, name the y-axis as Synchronous generator output voltage, and print the graph.

Refer to the user guide dealing with the computer-based instruments for EMS to know how to use the Graph window of the Metering application to obtain a graph, entitle a graph, name the axes of a graph, and print a graph.

Observe graph G621, which shows the voltage regulation characteristic of the synchronous generator when it supplies power to a resistive load. How does the output voltage E_O vary when the output current I_O increases? Briefly explain why.

___

___

___

In the Data Table window, clear the recorded data.

13. Replace the three-phase resistive load connected to the synchronous generator output (points A, B, and C in Figure 6-7) with the three-phase inductive load shown in Figure 6-8a. Make sure that all switches on the Inductive Load module are opened.

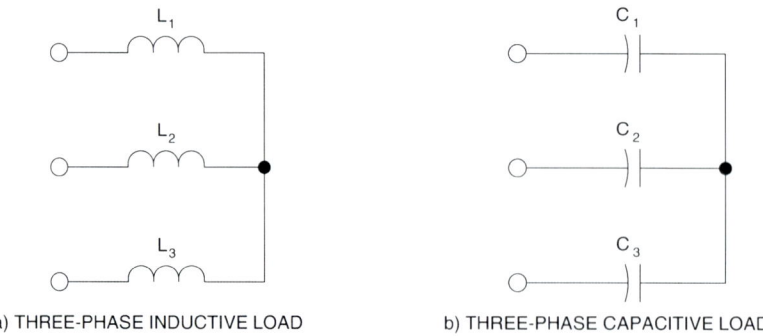

a) THREE-PHASE INDUCTIVE LOAD   b) THREE-PHASE CAPACITIVE LOAD

Figure 6-8. Three-phase inductive and capacitive loads.

14. Turn the Power Supply on and set the prime mover speed so that it rotates at the nominal speed of the Synchronous Motor/Generator.

    On the Synchronous Motor/Generator, set the EXCITER knob so that the line-to-line output voltage E_O of the synchronous generator is equal to the nominal value.

    Record the synchronous generator output voltage E_O, output current I_O, field current I_f, and speed n in the Data Table. These parameters are indicated by meters E line 1, I line 1, I field (I_f), and Speed, respectively.

15. Modify the settings on the load module so that the reactance X of the load decreases by steps as indicated in Table 6-5. You can refer to Appendix B of this manual to know how to obtain the various reactance values given in Table 6-5. For each reactance setting, readjust the prime mover speed so that it remains equal to the nominal speed of the Synchronous Motor/Generator, then record the data in the Data Table.

Table 6-5. Decreasing reactances X_1, X_2, and X_3 to load the synchronous generator.

Local ac power network		X_1, X_2, X_3 (Ω)	X_1, X_2, X_3 (Ω)	X_1, X_2, X_3 (Ω)	X_1, X_2, X_3 (Ω)	X_1, X_2, X_3 (Ω)	X_1, X_2, X_3 (Ω)	X_1, X_2, X_3 (Ω)
Voltage (V)	Frequency (Hz)							
120	60	1200	600	400	300	240	200	171
220	50	4400	2200	1467	1100	880	733	629
220	60	4400	2200	1467	1100	880	733	629
240	50	4800	2400	1600	1200	960	800	686

16. When all data has been recorded, set the prime mover speed to 0, then turn the Power Supply off.

17. In the Data Table window, confirm that the data has been stored, entitle the Data Table as DT622, and print the Data Table.

18. In the Graph window, make the appropriate settings to obtain a graph of the synchronous generator output voltage E_O (obtained from meter E line 1) as a function of the output current I_O (obtained from meter I line 1). Entitle the graph as G622, name the x-axis as Synchronous generator output current, name the y-axis as Synchronous generator output voltage, and print the graph.

19. Observe graph G622, which shows the voltage regulation characteristic of the synchronous generator when it supplies power to an inductive load. How does the output voltage E_O vary when the output current I_O increases?

_____

_____

Ex. 6-2 – Voltage Regulation Characteristics ♦ Procedure

Compare the voltage regulation characteristics obtained with the resistive load and the inductive load.

___

___

___

In the Data Table window, clear the recorded data.

20. Replace the three-phase inductive load connected to the synchronous generator output (points A, B, and C in Figure 6-7) with the three-phase capacitive load shown in Figure 6-8b. Make sure that all switches on the Capacitive Load module are opened.

    Repeat steps 14 to 18 of this exercise to obtain a graph of the output voltage E_O versus the output current I_O for the synchronous generator supplying power to a capacitive load. Entitle the Data Table and graph as DT623 and G623, respectively.

    Observe graph G623, which shows the voltage regulation characteristic of the synchronous generator when it supplies power to a capacitive load. How does the output voltage E_O vary when the output current I_O increases?

    ___

    ___

    Compare the voltage regulation characteristics of the synchronous generator (graphs G621 to G623) to those obtained with a single-phase transformer in Unit 7 of the student manual entitled *Power Circuits and Transformers*.

    ___

    ___

    ___

21. On the Power Supply, set the 24 V - AC power switch to the O (off) position.

     If you are using the Four-Quadrant Dynamometer/Power Supply, Model 8960-2, turn it off by setting its POWER INPUT switch to the O (off) position.

    Remove all leads and cables.

Ex. 6-2 – Voltage Regulation Characteristics ◆ *Conclusion*

CONCLUSION

In this exercise, you obtained the voltage regulation characteristics of a three-phase synchronous generator. You observed that the output voltage decreases as the output current increases when the synchronous generator supplies power to either a resistive or inductive load. You saw that the output voltage increases as the output current increases when the synchronous generator supplies power to a capacitive load. You found that the voltage regulation characteristics of the synchronous generator are similar to those of a single-phase transformer because the equivalent circuit is almost the same for both.

REVIEW QUESTIONS

1. The output voltage of a synchronous generator is a function of the

    a. speed of rotation and polarity of the field current.
    b. speed of rotation and strength of the field electromagnet.
    c. speed of rotation and input torque.
    d. speed of rotation only.

2. The equivalent circuit for one phase of a three-phase synchronous generator operating at constant speed and fixed field current is

    a. identical to that of a dc generator.
    b. very similar to that of a single-phase transformer.
    c. the same as a three-phase balanced circuit.
    d. the same as that of a dc battery.

3. In the equivalent circuit of the synchronous generator, reactance X_S is called

    a. stationary reactance.
    b. steady-state reactance.
    c. simplified reactance.
    d. synchronous reactance.

4. The voltage regulation characteristics of a synchronous generator are

    a. very similar to those of a single-phase transformer.
    b. quite different from those of a single-phase transformer.
    c. identical to those of a single-phase motor.
    d. only useful when the generator operates without load.

5. In the equivalent circuit of the synchronous generator, the value of X_S, expressed in ohms,

    a. is much smaller than the value of R_S.
    b. is much greater than the value of R_S.
    c. is the same value as R_S.
    d. depends on the generator output voltage.

Exercise 6-3

Frequency and Voltage Regulation

EXERCISE OBJECTIVE   When you have completed this exercise, you will be able to demonstrate frequency and voltage regulation of a synchronous generator using the Synchronous Motor/Generator module.

DISCUSSION   For a synchronous generator to operate as a power source that delivers a constant voltage at a fixed frequency, the speed of rotation and the strength of the field electromagnet must be controlled. As you saw in the previous exercise, resistive, inductive, and capacitive loads greatly affect the output voltage of a synchronous generator. Resistive loads also greatly affect the speed of a synchronous generator. However, inductive and reactive loads have little effect on the speed of rotation.

To obtain a constant output voltage and a fixed frequency from a synchronous generator under varying load conditions, the rotation speed and field current I_f must be adjusted simultaneously. In practice, automatic control systems continuously adjust the torque acting on the synchronous generator as well as the value of the field current I_f. For example, in hydroelectric systems, the torque is adjusted by changing the water turbine inlet size so as to maintain a constant speed, and thereby, a fixed frequency. The field current I_f is usually adjusted using power electronic devices so as to maintain a constant voltage. Manual adjustment of both the speed and field current at the same time is rather difficult to achieve, as you will observe in this exercise.

Procedure summary

In the first part of the exercise, you will set up the equipment in the Workstation, connect the equipment as shown in Figure 6-9, and make the appropriate settings on the equipment.

In the second part of the exercise, you will set the speed of rotation and the field current of the synchronous generator so that the frequency and output voltage are equal to the nominal values. You will change the nature of the load connected to the synchronous generator to observe how this affects the frequency and output voltage.

In the third part of the exercise, you will vary both the speed of rotation and the field current of the synchronous generator so as to maintain a constant output voltage and a fixed frequency under different load conditions.

EQUIPMENT REQUIRED

Refer to the Equipment Utilization Chart in Appendix C to obtain the list of equipment required for this exercise.

Ex. 6-3 – Frequency and Voltage Regulation ◆ *Procedure*

Procedure

High voltages are present in this laboratory exercise. Do not make or modify any banana jack connections with the power on unless otherwise specified.

Setting up the equipment

1. Install the equipment required in the EMS workstation.

   Mechanically couple the prime mover/dynamometer module to the Synchronous Motor/Generator.

2. On the Power Supply, make sure the main power switch is set to the O (off) position, and the voltage control knob is turned fully counterclockwise. Ensure the Power Supply is connected to a three-phase power source.

    If you are using the Four-Quadrant Dynamometer/Power Supply, Model 8960-2, connect its POWER INPUT to a wall receptacle.

3. Ensure that the data acquisition module is connected to a USB port of the computer.

   Connect the POWER INPUT of the data acquisition module to the 24 V - AC output of the Power Supply.

    If you are using the Prime Mover/Dynamometer, Model 8960-1, connect its LOW POWER INPUT to the 24 V - AC output of the Power Supply.

   On the Power Supply, set the 24 V - AC power switch to the I (on) position.

    If you are using the Four-Quadrant Dynamometer/Power Supply, Model 8960-2, turn it on by setting its POWER INPUT switch to the I (on) position. Press and hold the FUNCTION button 3 seconds to have uncorrected torque values on the display of the Four-Quadrant Dynamometer/Power Supply. The indication "NC" appears next to the function name on the display to indicate that the torque values are uncorrected.

4. Start the Data Acquisition software (LVDAC or LVDAM). Open setup configuration file ACMOTOR1.DAI.

   In the Metering window, select layout 2. Make sure that the continuous refresh mode is selected.

5. Connect the equipment as shown in Figure 6-9. Open all switches on the Resistive, Inductive, and Capacitive Load modules.

   On the Synchronous Motor/Generator, set the EXCITER switch to the I (closed) position and the EXCITER knob to the mid position.

Ex. 6-3 – Frequency and Voltage Regulation ♦ *Procedure*

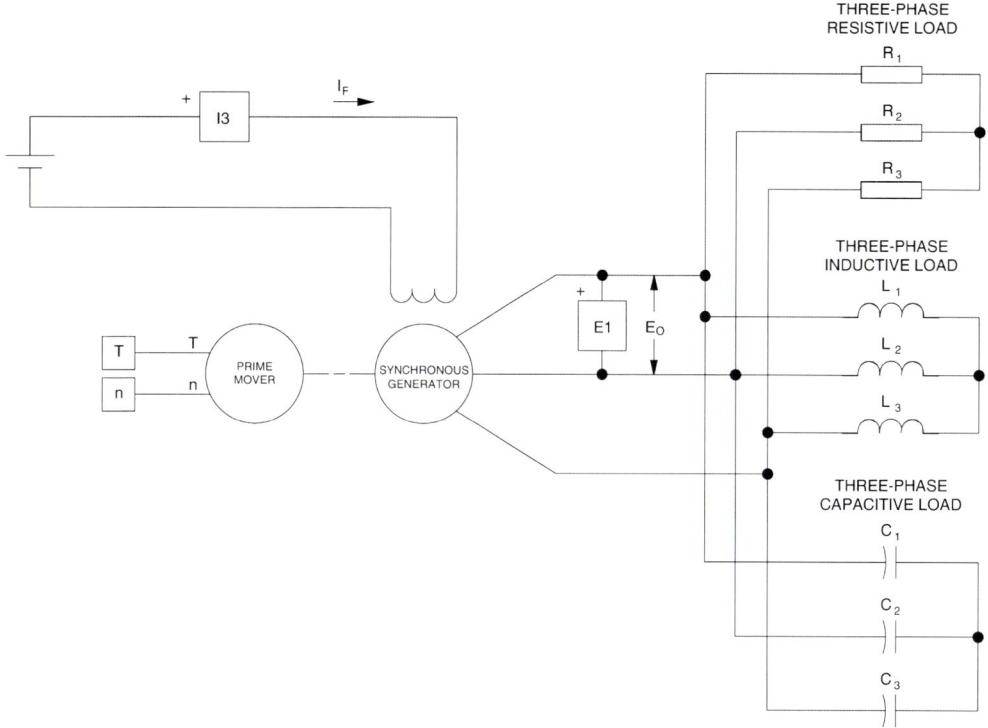

Figure 6-9. Synchronous generator under load coupled to a prime mover.

6. Set the Four-Quadrant Dynamometer/Power Supply or the Prime Mover/Dynamometer to operate as a clockwise prime mover. To do this, refer to Exercise 1-1 or Exercise 1-2 if necessary.

    If you are performing the exercise using LVSIM®-EMS, you can zoom in the Prime Mover/Dynamometer module before setting the controls in order to see additional front panel markings related to these controls.

Effect of the load on the output voltage and the frequency

7. Turn the Power Supply on. Set the prime mover speed at the nominal speed of the Synchronous Motor/Generator.

    The rating of any of the machines is indicated in the lower left corner of the module front panel.

Ex. 6-3 – Frequency and Voltage Regulation ◆ *Procedure*

8. On the Synchronous Motor/Generator, set the EXCITER knob so that the line-to-line output voltage E_O of the synchronous generator (indicated by meter E line 1 in the Metering window) is equal to the nominal value.

   Record the nominal output voltage E_O and frequency f in the following blank spaces. These parameters are indicated by meters E line 1 and Frequency, respectively.

   E_O (nominal) = _____ V

   f (nominal) = _____ Hz

9. On the Resistive Load module, set the resistance of resistors R_1, R_2, and R_3 to the value indicated in the following table.

   Table 6-6. Load value.

Local ac power network		Resistance or reactance (Ω)
Voltage (V)	Frequency (Hz)	
120	60	240
220	50	880
220	60	880
240	50	960

   Record the output voltage E_O and frequency f in the following blank spaces.

   E_O = _____ V (resistive load)

   f = _____ Hz (resistive load)

   How do the output voltage and frequency vary when a resistive load is connected to the synchronous generator output?

   _____

   _____

   Open all switches on the Resistive Load module. Wait for the frequency and output voltage to stabilize. They should be equal to the nominal values.

Ex. 6-3 – Frequency and Voltage Regulation ◆ *Procedure*

10. On the Inductive Load module, set the reactance of inductors L_1, L_2, and L_3 to the value indicated in Table 6-6.

    Record the output voltage E_O and frequency f in the following blank spaces.

    E_O = _____ V (inductive load)

    f = _____ Hz (inductive load)

    How do the output voltage and frequency vary when an inductive load is connected to the synchronous generator output?

    _____

    _____

    Open all switches on the Inductive Load module. Wait for the frequency and output voltage to stabilize. They should be equal to the nominal values.

11. On the Capacitive Load module, set the reactance of capacitors C_1, C_2, and C_3 to the value indicated in Table 6-6.

    Record the output voltage E_O and frequency f in the following blank spaces.

    E_O = _____ V (capacitive load)

    f = _____ Hz (capacitive load)

    How do the output voltage and frequency vary when a capacitive load is connected to the synchronous generator output?

    _____

    _____

    Open all switches on the Capacitive Load module. Wait for the frequency and output voltage to stabilize. They should be equal to the nominal values.

12. Compare the effect of the resistive, inductive, and capacitive loads on the synchronous generator output voltage.

    _____

    _____

Ex. 6-3 – Frequency and Voltage Regulation ◆ *Procedure*

Compare the effect of the resistive, inductive, and capacitive loads on the frequency of the voltages produced by the synchronous generator.

_____

_____

_____

Frequency and voltage regulation

13. On the Inductive Load module, set the reactance of inductors L_1, L_2, and L_3 to the value indicated in Table 6-7.

Table 6-7. Reactance of inductors L_1, L_2, and L_3.

Local ac power network		Reactance of L_1, L_2, and L_3 (Ω)
Voltage (V)	Frequency (Hz)	
120	60	600
220	50	1467
220	60	1467
240	50	1200

Readjust the prime mover speed and the EXCITER knob of the Synchronous Motor/Generator so that the synchronous generator output voltage and frequency are equal to the nominal values.

14. On the Capacitive Load module, set the reactance of capacitors C_1, C_2, and C_3 to the value indicated in Table 6-8.

Table 6-8. Reactance of capacitors C_1, C_2, and C_3.

Local ac power network		Reactance of C_1, C_2, and C_3 (Ω)
Voltage (V)	Frequency (Hz)	
120	60	300
220	50	2200
220	60	2200
240	50	2400

Readjust the prime mover speed and the EXCITER knob of the Synchronous Motor/Generator so that the synchronous generator output voltage and frequency are equal to the nominal values.

Ex. 6-3 – Frequency and Voltage Regulation ♦ *Conclusion*

15. On the Resistive Load module, set the resistance of resistors R_1, R_2, and R_3 to the value indicated in Table 6-9.

Table 6-9. Resistance of resistors R_1, R_2, and R_3.

Local ac power network		Resistance of R_1, R_2, and R_3 (Ω)
Voltage (V)	Frequency (Hz)	
120	60	200
220	50	880
220	60	880
240	50	800

Readjust the prime mover speed and the EXCITER knob of the Synchronous Motor/Generator so that the synchronous generator output voltage and frequency are equal to the nominal values.

Is it easy to rapidly readjust the synchronous generator output voltage and frequency when the load changes? Why?

_____

_____

_____

16. On the Power Supply, set the 24 V - AC power switch to the O (off) position.

 If you are using the Four-Quadrant Dynamometer/Power Supply, Model 8960-2, turn it off by setting its POWER INPUT switch to the O (off) position.

Remove all leads and cables.

CONCLUSION

In this exercise, you observed that the output voltage and frequency of a synchronous generator change whether a resistive, inducti.ve, or capacitive load is connected to the output. You observed that resistive loads have a greater effect on frequency than inductive and capacitive loads. You found that maintaining the frequency and output voltage to the nominal values, when the load changes, is rather difficult to achieve manually. This is because both the speed of rotation and the field current of the synchronous generator must be adjusted to correct the changes in frequency and voltage.

Ex. 6-3 – Frequency and Voltage Regulation ◆ *Review Questions*

REVIEW QUESTIONS

1. When the load connected to a synchronous generator changes,

    a. there is no effect on the output voltage nor the frequency.
    b. both the output voltage and the frequency are affected.
    c. only the output voltage is affected.
    d. only the frequency is affected.

2. For a synchronous generator to deliver a constant output voltage at a fixed frequency,

    a. both its speed and field current must be controlled.
    b. only its speed must be controlled.
    c. only its excitation current must be controlled.
    d. the load must only be resistive.

3. Manual adjustment of the speed and field current to maintain the output voltage and frequency of a synchronous generator to the nominal values is

    a. a simple task.
    b. a rather difficult task.
    c. only possible when the synchronous generator is fully loaded.
    d. only possible when the synchronous generator is exactly at half load.

4. Inductive and capacitive loads have little effect on

    a. both the output voltage and frequency of a synchronous generator.
    b. the frequency of a synchronous generator.
    c. the output voltage of a synchronous generator.
    d. the nominal power rating of a synchronous generator.

5. Resistive loads have a great effect on

    a. both the output voltage and frequency of a synchronous generator.
    b. only the frequency of a synchronous generator.
    c. only the output voltage of a synchronous generator.
    d. the nominal power rating of a synchronous generator.

Exercise 6-4

Generator Synchronization

EXERCISE OBJECTIVE When you have completed this exercise, you will be able to synchronize a three-phase synchronous generator with the ac power network using the Synchronous Motor/Generator and the Synchronizing Module.

DISCUSSION Most of the electricity consumed today is produced by three-phase synchronous generators. Since a huge amount of electricity is consumed every day, ac power networks are generally made up of a large number of synchronous generators all operating at the same frequency. When the power demand increases, additional generators are connected to the ac power network. Before connecting a three-phase synchronous generator to an ac power network, the following conditions are to be observed:

- The frequency of the voltages produced by the generator must be equal to the ac power network frequency.

- The value of the voltages produced by the generator must be equal to the ac power network voltage.

- The phase sequence of the voltages produced by the generator must be the same as that of the ac power network.

- The voltages produced by the generator must be in phase with the ac power network voltages.

A generator is said to be synchronized when all these conditions are met. A synchronous generator must never be connected to an ac power network before verifying synchronization. Connecting a non-synchronized generator to an ac power network could cause severe damage to the generator, because of the high torque that would be applied to the generator's shaft and the huge currents that would flow in the generator windings at connection.

Once a synchronous generator is connected to an ac power network, no current flows between the generator and the ac power network because they produce voltages having the same amplitude and phase. As a result, the generator supplies neither active nor reactive power to the ac power network. In this case, the generator is said to be "floating" on the ac power network. Furthermore, its frequency can no longer be changed by adjusting the torque applied to the generator's shaft. This is because the ac power network is so powerful that it imposes its own frequency. However, adjusting the torque applied to the generator's shaft allows changing the amount of active power that is exchanged between the generator and the ac power network. Increasing the torque increases the amount of active power that is delivered to the ac power network. Conversely, decreasing the torque decreases the amount of active power that is delivered to the ac power network. The generator could even receive active power from the ac power network, and thus operate as a synchronous motor, if the torque applied to the generator's shaft were decreased to zero.

Ex. 6-4 – Generator Synchronization ◆ *Discussion*

As in three-phase synchronous motors, the amount of reactive power that is exchanged between a synchronous generator and the ac power network can be changed by adjusting the field current. The field current is usually adjusted so that no reactive power is exchanged between the generator and the ac power network, i.e., so that the power factor of the generator is unity. This minimizes the line currents and allows the size of the conductors connecting the generator to the ac power network to be reduced to minimum.

Figure 6-10 shows a simple circuit used to synchronize and connect a generator and an ac power network. In this circuit, a three-phase synchronous generator is connected to a three-phase power network (three-phase power source) through three lamps and a three-pole switch set to the open position. A voltmeter and a frequency meter are connected to the generator output to measure its voltage and frequency.

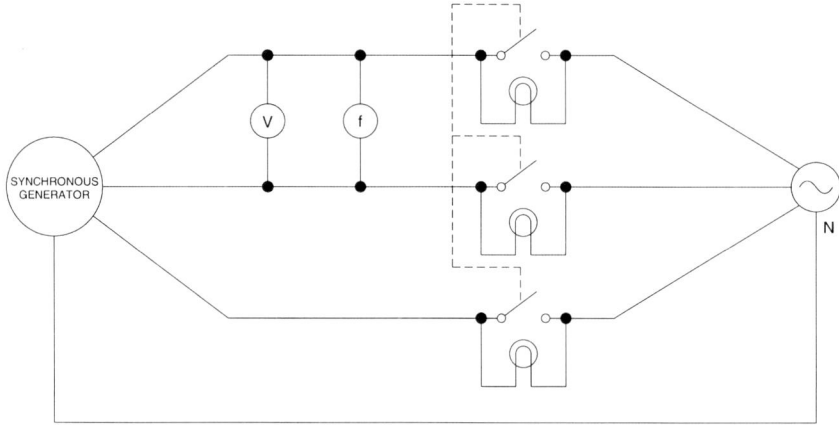

Figure 6-10. Circuit used to synchronize and connect a generator and an ac power network.

The speed and field current of the synchronous generator are first adjusted so that the generator frequency and voltage are approximately equal to the nominal voltage and frequency of the ac power network. The brightness of the lamps will change in synchronism when the phase sequence of the generator is the same as that of the ac power network. On the other hand, the lamp brightness will change out of synchronism if the phase sequence of the generator differs from that of the ac power network. In this case, the connections of two of the three line wires of the synchronous generator must be interchanged to reverse its phase sequence.

Once the phase sequence of the synchronous generator is correct, the speed of the generator is adjusted so that the rate at which the lamp brightness changes is as low as possible. This adjusts the frequency of the generator to that of the ac power network. The field current of the generator is then adjusted so that the lamps become completely dimmed as their brightness decreases. This adjusts the generator voltage to that of the ac power network. The switch can then be closed at any instant the lamps are dimmed completely (the voltages are in phase at this instant only) to safely connect the synchronous generator to the ac power network.

Ex. 6-4 – Generator Synchronization ♦ *Procedure*

Procedure summary

In the first part of the exercise, you will set up the equipment in the Workstation, connect the equipment as shown in Figure 6-11, and make the appropriate settings on the equipment.

In the second part of the exercise, you will synchronize a three-phase synchronous generator with the three-phase power network. You will then connect the synchronous generator to the three-phase power network.

In the third part of the exercise, you will vary the torque applied to the generator's shaft and the field current I_f and observe how this affects the operation of the synchronous generator.

EQUIPMENT REQUIRED

Refer to the Equipment Utilization Chart in Appendix C to obtain the list of equipment required for this exercise.

PROCEDURE

 High voltages are present in this laboratory exercise. Do not make or modify any banana jack connections with the power on unless otherwise specified.

Setting up the equipment

1. Install the equipment required in the EMS workstation.

   Mechanically couple the prime mover/dynamometer module to the Synchronous Motor/Generator.

2. On the Power Supply, make sure the main power switch is set to the O (off) position, and the voltage control knob is turned fully counterclockwise. Ensure the Power Supply is connected to a three-phase power source.

    If you are using the Four-Quadrant Dynamometer/Power Supply, Model 8960-2, connect its POWER INPUT to a wall receptacle.

3. Ensure that the data acquisition module is connected to a USB of the computer.

   Connect the POWER INPUT of the data acquisition module to the 24 V - AC output of the Power Supply.

    If you are using the Prime Mover/Dynamometer, Model 8960-1, connect its LOW POWER INPUT to the 24 V - AC output of the Power Supply.

Ex. 6-4 – Generator Synchronization ♦ *Procedure*

On the Power Supply, set the 24 V - AC power switch to the I (on) position

 If you are using the Four-Quadrant Dynamometer/Power Supply, Model 8960-2, turn it on by setting its POWER INPUT switch to the I (on) position. Press and hold the FUNCTION button 3 seconds to have uncorrected torque values on the display of the Four-Quadrant Dynamometer/Power Supply. The indication "NC" appears next to the function name on the display to indicate that the torque values are uncorrected.

4. Start the Data Acquisition software (LVDAC or LVDAM). Open setup configuration file ACMOTOR1.DAI.

   In the Metering window, select layout 2. Make sure that the continuous refresh mode is selected.

5. Connect the equipment as shown in Figure 6-11.

   On the Synchronous Motor/Generator, set the EXCITER switch to the I (close) position and the EXCITER knob to the mid position.

   On the Synchronizing Module, set the switch to the O (open) position

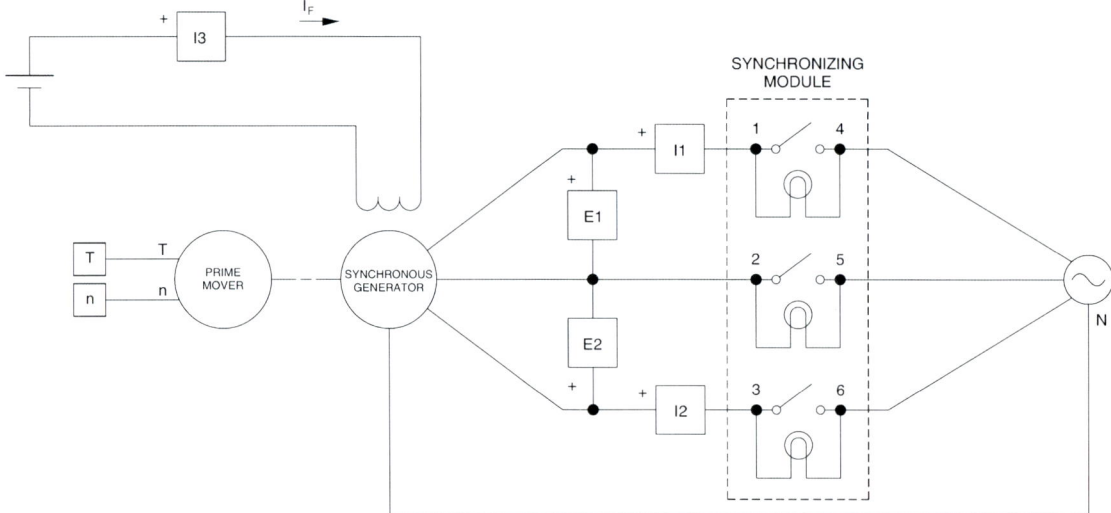

Figure 6-11. Circuit used to synchronize and connect a generator and an ac power network.

6. Set the Four-Quadrant Dynamometer/Power Supply or the Prime Mover/Dynamometer to operate as a clockwise prime mover.

 If you are performing the exercise using LVSIM®-EMS, you can zoom in the Prime Mover/Dynamometer module before setting the controls in order to see additional front panel markings related to these controls.

Generator synchronization

7. On the Synchronous Motor/Generator, interchange the connections of the leads at terminals 1 and 2.

   Set the prime mover speed at the nominal speed of the Synchronous Motor/Generator minus approximately 75 r/min.

    The rating of any of the machines is indicated in the lower left corner of the module front panel.

8. On the Synchronous Motor/Generator, set the EXCITER knob so that the line-to-line output voltage E_O of the synchronous generator (indicated by meter E line 1 in the Metering window) is equal to the nominal value.

   Observe the lamps on the Synchronizing Module.

   Does the phase sequence of the synchronous generator correspond to that of the three-phase power network? Why?

   _____

9. Leave the prime mover speed set as it is. Turn the Power Supply off.

   On the Synchronous Motor/Generator, interchange the connections of the leads at terminals 1 and 2.

10. Turn the Power Supply on.

    Observe the lamps on the Synchronizing Module.

    Does the phase sequence of the synchronous generator correspond to that of the three-phase power network? Why?

    _____

11. Set the prime mover speed so that the brightness of the lamps on the Synchronizing Module changes very slowly (if necessary).

    Is the generator synchronized with the three-phase power network at the instants the lamps are dimmed completely?

    ❏ Yes   ❏ No

12. On the Synchronizing Module, set the switch to the I (closed) position at an instant the lamps are dimmed completely. This connects the synchronous generator to the three-phase power network.

Ex. 6-4 – Generator Synchronization ♦ *Procedure*

In the Metering window, observe the active power indicated by meter React. power. Is a significant amount of active power exchanged between the synchronous generator and the ac power network?

❏ Yes   ❏ No

Effect of the torque and field current on the generator operation

13. In the Metering window, make sure that the torque correction function of the Torque meter is enabled.

Slowly increase the prime-mover speed setting to increase the generator input torque until the Torque meter reads -1.0 N·m (9.0 lbf·in). While doing this, observe the active power and the generator speed indicated by meters Act. power and Speed.

The synchronous generator delivers active power when the value indicated by meter Act. power is positive.

Describe what happens.

_____

_____

_____

Does the synchronous generator supply active power to the ac power network?

❏ Yes   ❏ No

14. Slowly set the prime-mover speed setting so that the active power indicated by meter Act. power decreases to approximately zero. While doing this, observe the generator input torque indicated by the Torque meter.

The synchronous generator is now "floating" on the ac power network. Where does the power to overcome friction come from?

_____

_____

15. Slowly decrease the prime-mover speed setting to 0. While doing this, observe the active power, the generator input torque, and the generator speed indicated by meters Act. power, Torque, and Speed.

Describe what happens.

_____

_____

_____

Ex. 6-4 – Generator Synchronization ◆ *Procedure*

What does this mean?

_____

_____

16. Increase the prime-mover speed setting until the torque indicated by the Torque meter (generator input torque) is equal to −1.0 N·m (9.0 lbf·in). The synchronous generator is now delivering the nominal active power (approximately) to the ac power network.

    On the Synchronous Motor/Generator, slowly set the EXCITER knob to the MAX. position to increase the field current. While doing this, observe the active power, the reactive power, the generator input torque, and the generator speed indicated by the meters.

    The synchronous generator delivers reactive power when the value indicated by meter React. power is positive.

    Describe what happens.

    _____

    _____

    _____

    Does the synchronous generator supply reactive power to the ac power network?

    ❑ Yes   ❑ No

17. On the Synchronous Motor/Generator, slowly set the EXCITER knob to the MIN. position to decrease the field current. While doing this, observe the reactive power indicated by meter React. power in the Metering window.

    Describe what happens.

    _____

    _____

    _____

    Is it possible to adjust the field current so that the power factor of the synchronous generator is unity?

    ❑ Yes   ❑ No

Ex. 6-4 – Generator Synchronization ♦ Conclusion

18. On the Power Supply, set the 24 V - AC power switch to the O (off) position.

 If you are using the Four-Quadrant Dynamometer/Power Supply, Model 8960-2, turn it off by setting its POWER INPUT switch to the O (off) position.

Remove all leads and cables.

CONCLUSION

In this exercise, you synchronized a three-phase synchronous generator with the ac power network. You observed that varying the torque at the generator's shaft varies the amount of active power exchanged between the generator and the ac power network. You saw that varying the field current of the generator varies the amount of reactive power exchanged between the generator and the ac power network.

REVIEW QUESTIONS

1. Before a synchronous generator is synchronized with the ac power network, its phase sequence, frequency, and voltage must be

   a. the same as those of the ac power network.
   b. different from those of the ac power network.
   c. any value depending on the generator and its prime mover.
   d. none of the above.

2. After synchronization with the ac power network, the phase sequence, frequency, and voltage of a synchronous generator will be

   a. the same as those of the ac power network.
   b. different from those of the ac power network.
   c. any value depending on the generator and its prime mover.
   d. none of the above.

3. What parameters of the synchronous generator must be adjusted before connecting it to an ac power network?

   a. Its phase sequence and frequency only.
   b. Its voltage and frequency only.
   c. Its phase sequence, frequency, and voltage.
   d. Its speed only.

4. When a synchronous generator "floats" on the ac power network, this means that

   a. it will speed up and slow down with network voltage fluctuations.
   b. neither active nor reactive power is exchanged with the ac power network.
   c. it is sitting above the water line.
   d. the output voltage is almost identical to that of the ac power network.

Ex. 6-4 – Generator Synchronization ◆ *Review Questions*

5. Active power to overcome the rotation friction of a synchronous generator that is "floating" on the ac power network comes from

   a. the network.
   b. the ac power supply.
   c. the source of mechanical power coupled to the generator.
   d. the field current.

Unit Test

1. A three-phase synchronous generator with thirty pairs of poles per stator windings produces voltages at a frequency of 60 Hz when it rotates at the nominal speed. What is the nominal speed of the synchronous generator?

   a. 3600 r/min
   b. 360 r/min
   c. 120 r/min
   d. 1800 r/min

2. A three-phase synchronous generator

   a. is very similar to an eddy-current brake.
   b. is basically an electromagnet that induces voltages in the stator windings as it rotates.
   c. consists of three electromagnets, located at 120° from each other, that induce voltages in the stator windings as they rotate.
   d. None of the above.

3. Changing the speed of a synchronous generator changes

   a. the frequency and amplitude of the output voltage.
   b. only the frequency of the output voltage.
   c. only the amplitude of the output voltage.
   d. only the phase of the output voltage.

4. How can the output voltage of a synchronous generator be varied without modifying the frequency?

   a. By varying the speed of the generator.
   b. By varying the way the conductors are wound in the stator winding.
   c. By varying the torque applied to the generator's shaft.
   d. By varying the field current of the generator.

5. When a synchronous generator supplies power to either a resistive or an inductive load, the output voltage

   a. decreases as the field current increases.
   b. increases as the output current increases.
   c. remains constant when the output current varies.
   d. decreases as the output current increases.

6. When a synchronous generator supplies power to a capacitive load, the output voltage

   a. increases as the field current decreases.
   b. increases as the output current increases.
   c. remains constant when the output current varies.
   d. decreases as the output current increases.

7. When a single synchronous generator supplies power to a load that fluctuates, the voltage and frequency can be maintained constant by continually adjusting the

   a. generator speed only.
   b. position of the stator windings.
   c. speed and field current of the generator.
   d. field current only.

8. A three-phase synchronous generator is said to be synchronized with the ac power network when

   a. its phase sequence is the same as that of the network.
   b. the amplitude and frequency of the voltages produced by the generator are the same as those of the network.
   c. the amplitude, frequency, and phase of the generator voltages are the same as those of the network.
   d. Both a and c.

9. When a synchronous generator is synchronized with the ac power network, increasing the torque applied to the generator's shaft

   a. increases the reactive power which the generator delivers.
   b. increases the active power which the generator delivers.
   c. decreases the reactive power which the generator delivers.
   d. increases the generator speed.

10. When a synchronous generator is synchronized with the ac power network, varying the field current changes

    a. the amount of active power which the generator delivers.
    b. the generator speed.
    c. the ac power network voltage.
    d. the power factor of the generator.

Appendix A

Circuit Diagram Symbols

Various symbols are used in many of the circuit diagrams given in the DISCUSSION and PROCEDURE sections of this manual. Each symbol is a functional representation of a device used in electrical power technology. For example, different symbols represent a fixed-voltage dc power supply, a variable-voltage single-phase ac power supply, and a synchronous motor/generator. The use of these symbols greatly simplifies the circuit diagrams, by reducing the number of interconnections shown, and makes it easier to understand operation.

For each symbol used in this and other manuals of the Electrical Power Technology using Data Acquisition series, this appendix gives the name of the device which the symbol represents and a diagram showing the equipment, and in some cases the connections, required to obtain the device. Notice that the terminals of each symbol are identified using encircled numbers. Identical encircled numbers identify the corresponding terminals in the equipment and connections diagram.

Symbol	Equipment and connections
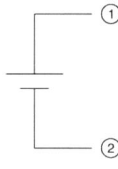 Fixed-voltage dc power supply	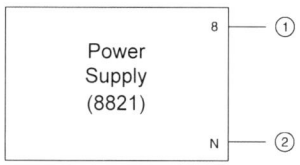
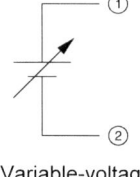 Variable-voltage dc power supply	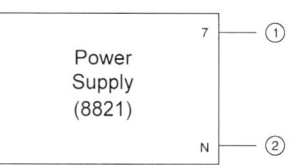

Appendix A   Circuit Diagram Symbols

Symbol

Equipment and connections

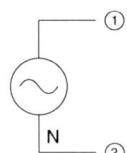

Fixed-voltage
ac power supply

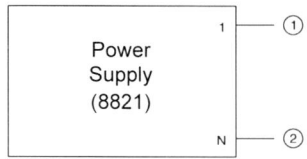

* Note: terminal 2 or 3
can also be used.

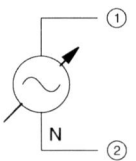

Variable-voltage
ac power supply

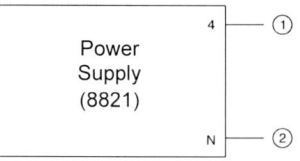

* Note: terminal 5 or 6
can also be used.

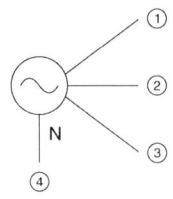

Fixed-voltage three-phase
ac power supply

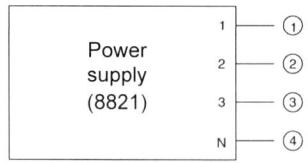

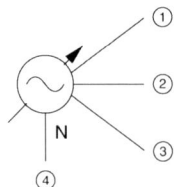

Variable-voltage three-phase
ac power supply

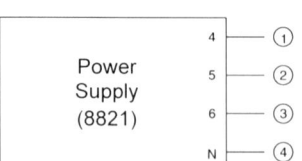

278   © Festo Didactic 584108

Appendix A

Circuit Diagram Symbols

Symbol   **Equipment and connections**

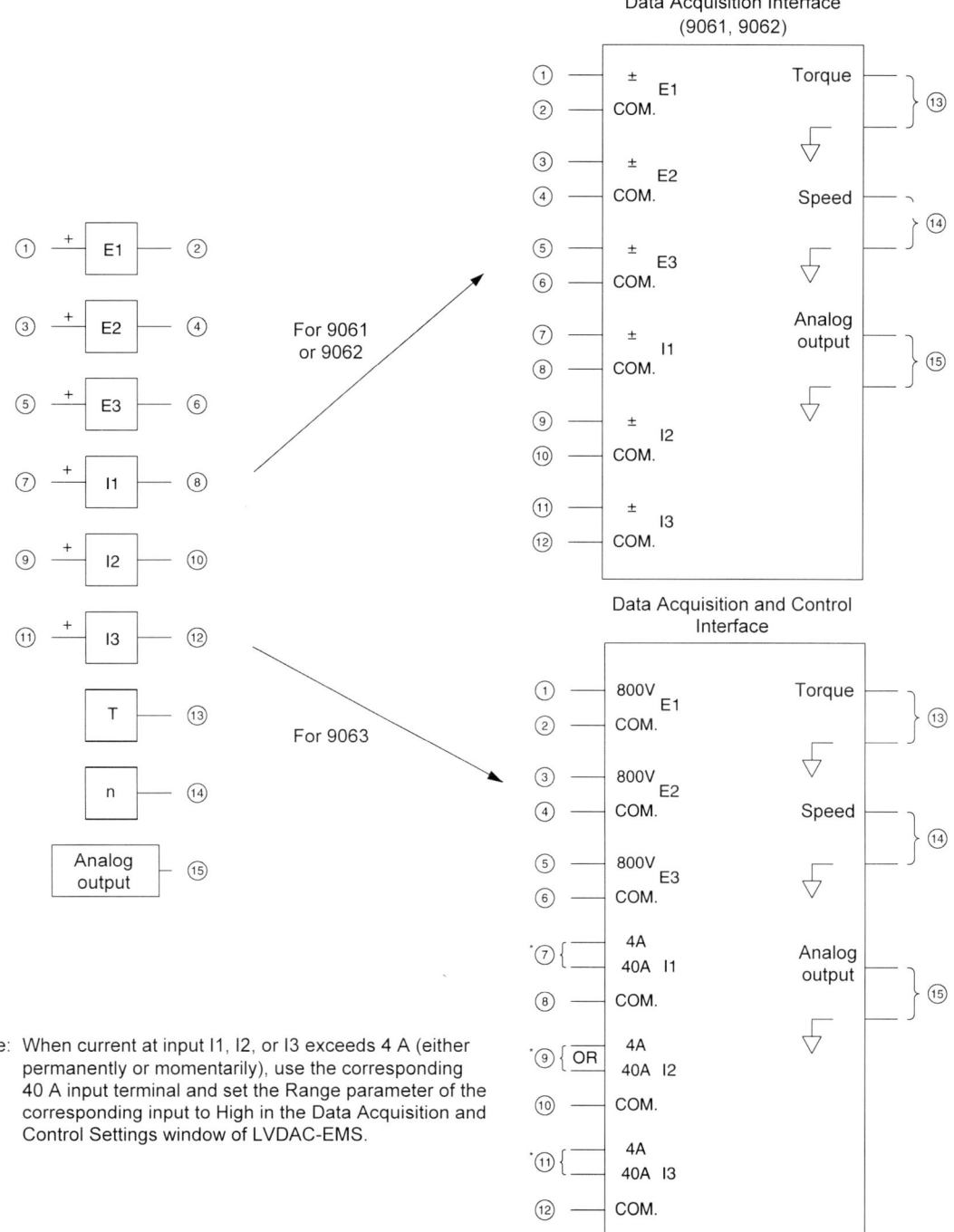

*Note: When current at input I1, I2, or I3 exceeds 4 A (either permanently or momentarily), use the corresponding 40 A input terminal and set the Range parameter of the corresponding input to High in the Data Acquisition and Control Settings window of LVDAC-EMS.

Appendix A

Circuit Diagram Symbols

Symbol

Equipment and connections

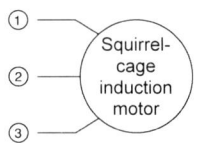

Squirrel-cage
induction motor

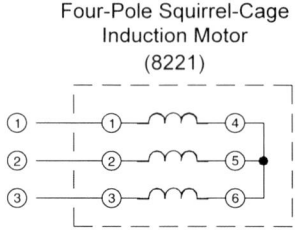

Four-Pole Squirrel-Cage
Induction Motor
(8221)

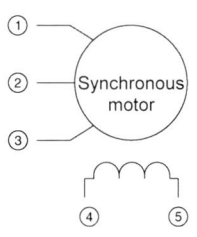

Synchronous
motor

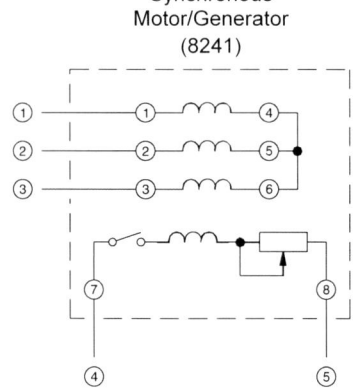

Synchronous
Motor/Generator
(8241)

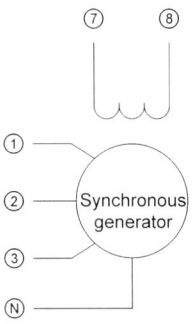

Three-phase
synchronous generator

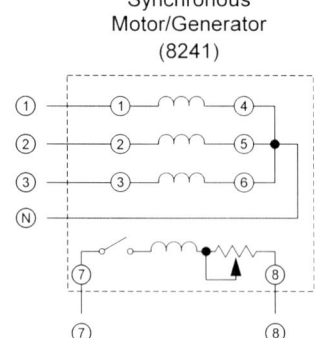

Synchronous
Motor/Generator
(8241)

Appendix A   Circuit Diagram Symbols

Symbol   **Equipment and connections**

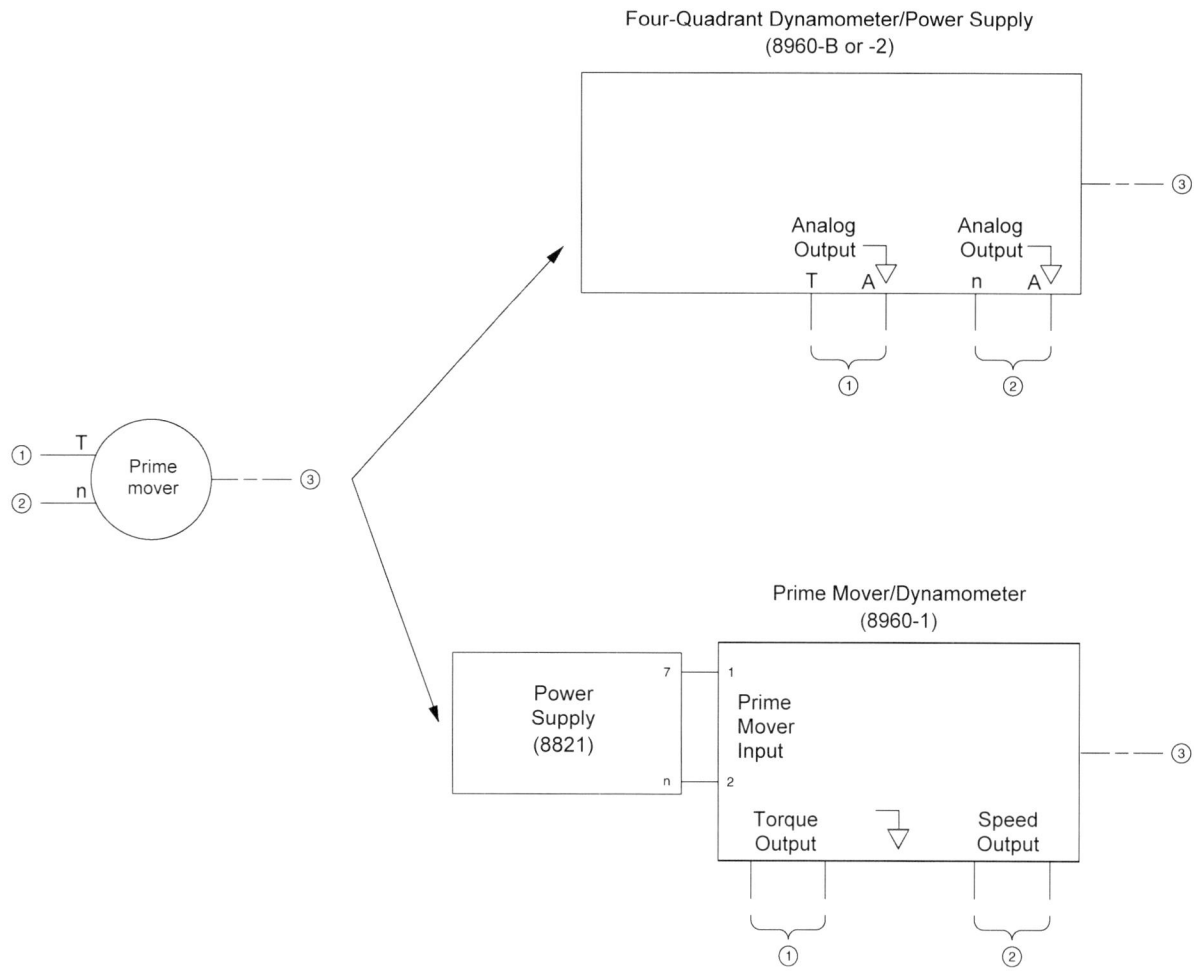

Appendix A Circuit Diagram Symbols

Symbol   **Equipment and connections**

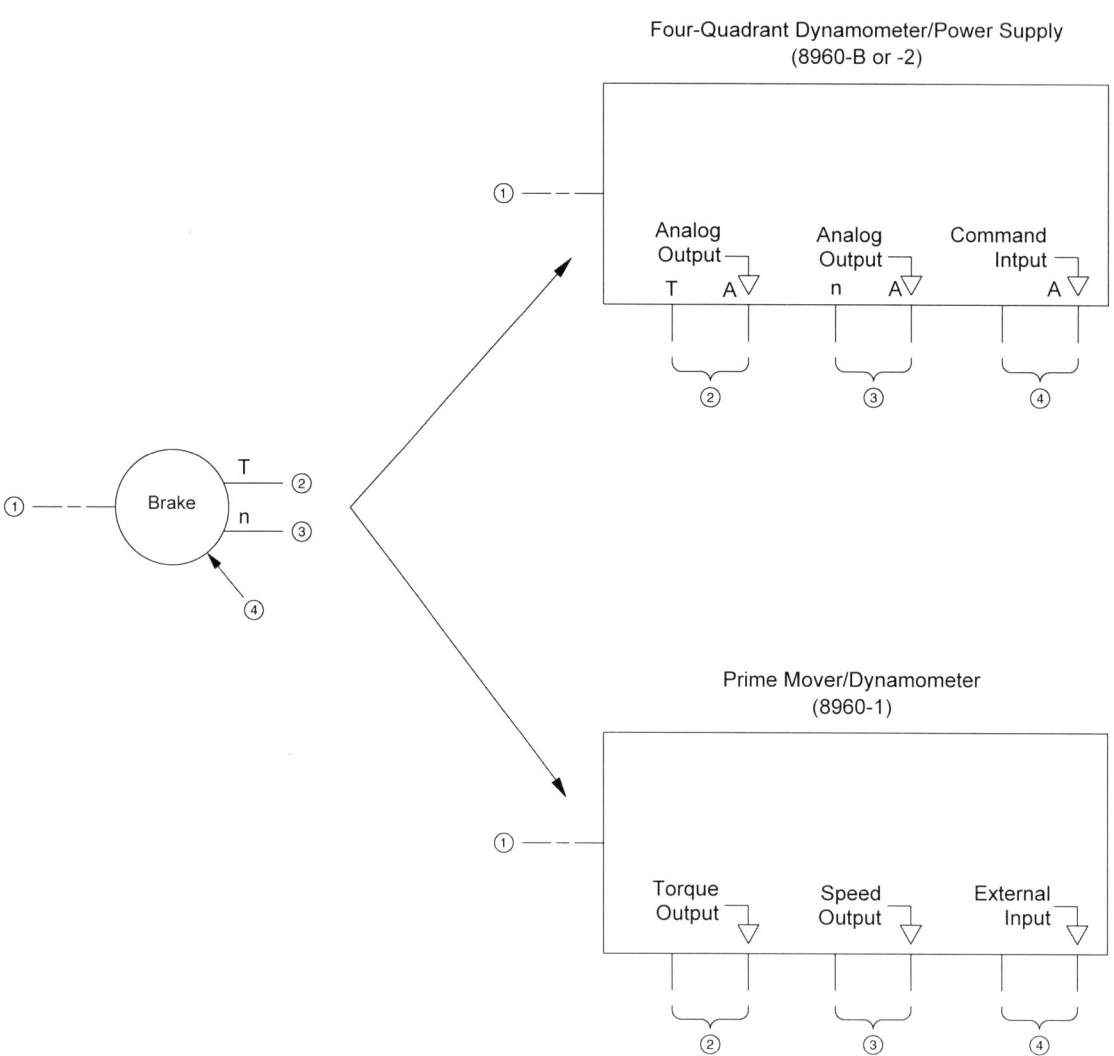

Appendix B

Impedance Table for the Load Modules

The following table gives impedance values which can be obtained using either the Resistive Load, Model 8311, the Inductive Load, Model 8321, or the Capacitive Load, Model 8331. Figure B-1 shows the load elements and connections. Other parallel combinations can be used to obtain the same impedance values listed.

Table B-1. Impedance table for the load modules.

Impedance (Ω)			Position of the switches								
120 V 60 Hz	220 V 50 Hz/60 Hz	240 V 50 Hz	1	2	3	4	5	6	7	8	9
1200	4400	4800	I								
600	2200	2400		I							
300	1100	1200			I						
400	1467	1600	I	I							
240	880	960	I		I						
200	733	800		I	I						
171	629	686	I	I	I						
150	550	600	I			I	I	I			
133	489	533		I		I	I	I			
120	440	480			I		I	I			
109	400	436			I	I	I	I			
100	367	400	I		I	I	I	I			
92	338	369		I	I	I	I	I			
86	314	343	I	I	I	I	I	I			
80	293	320	I			I	I	I	I	I	I
75	275	300		I		I	I	I	I	I	I
71	259	282			I		I	I	I	I	I
67	244	267			I	I	I	I	I	I	I
63	232	253	I		I	I	I	I	I	I	I
60	220	240		I	I	I	I	I	I	I	I
57	210	229	I	I	I	I	I	I	I	I	I

Appendix B

Impedance Table for the Load Modules

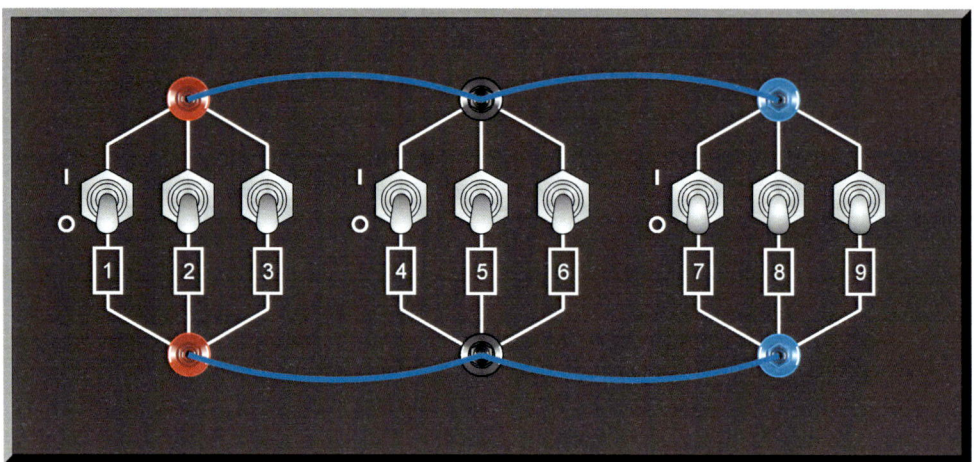

Figure B-1. Location of the load elements on the Resistive Load, Inductive Load, and Capacitive Load, Models 8311, 8321, and 8331, respectively.

Appendix B — Impedance Table for the Load Modules

The following table gives inductance values which can be obtained using the Inductive Load module, Model 8321. Figure B-1 shows the load elements and connections. Other parallel combinations can be used to obtain the same inductance values listed.

Table B-2. Inductance table for the Inductive Load module.

Inductance (H)				Position of the switches								
120 V 60 Hz	220 V 50 Hz	220 V 60 Hz	240 V 50 Hz	1	2	3	4	5	6	7	8	9
3.20	14.00	11.70	15.30	I								
1.60	7.00	5.80	7.60		I							
1.07	4.67	2.90	5.08	I	I							
0.80	3.50	3.88	3.80			I						
0.64	2.80	2.32	3.04	I		I						
0.53	2.33	1.93	2.53		I	I						
0.46	2.00	1.66	2.17	I	I	I						
0.40	1.75	1.45	1.90	I			I	I	I			
0.36	1.56	1.29	1.69		I		I	I	I			
0.32	1.40	1.16	1.52			I		I	I			
0.29	1.27	1.06	1.38			I	I	I	I			
0.27	1.17	0.97	1.27	I		I	I	I	I			
0.25	1.08	0.89	1.17		I	I	I	I	I			
0.23	1.00	0.83	1.09	I	I	I	I	I	I			
0.21	0.93	0.77	1.01	I			I	I	I	I	I	I
0.20	0.88	0.73	0.95		I		I	I	I	I	I	I
0.19	0.82	0.68	0.89			I		I	I	I	I	I
0.18	0.78	0.65	0.85			I	I	I	I	I	I	I
0.17	0.74	0.61	0.80	I		I	I	I	I	I	I	I
0.16	0.70	0.58	0.76		I	I	I	I	I	I	I	I
0.15	0.67	0.55	0.72	I	I	I	I	I	I	I	I	I

Appendix C

Equipment Utilization Chart

The following equipment is required to perform the exercises in this manual.

Equipment		Exercise							
Model	Description	1-1	1-2	1-3	2-1	2-2	2-3	3-1	3-2
8134[1]	EMS Workstation	1	1	1	1	1	1	1	1
8211	DC Motor/Generator	1	1	1	1	1	1	1	1
8221[2]	Four-Pole Squirrel-Cage Induction Motor	1	1						
8241	Three-Phase Synchronous Motor/Generator								
8251	Capacitor-Start Motor								
8254	Universal Motor								1
8311	Resistive Load					1[3]	1	1	
8321	Inductive Load								
8331	Capacitive Load								
8621	Synchronizing Module								
8821-2X	Power Supply	1	1	1	1	1	1	1	1
8942	Timing Belt	1	1	1	1	1	1	1	1
8951	Connection Leads and Accessories	1	1	1	1	1	1	1	1
8960-1 or -2	Prime Mover and Dynamometer Module	1[4]	1[5]	1	1	1	1	1	1
9061, 9062, or 9063	Data Acquisition Module	1	1	1	1	1	1	1	1

[1] Workstation model 8110-2 can also be used.

[2] Model 8221-2 can also be used.

[3] Not required when performing the exercise using 120-V equipment.

[4] Exercise 1-1 can be performed only with the Four-Quadrant Dynamometer/Power Supply, Model 8960-2 (or 8960-B).

[5] Exercise 1-2 can be performed only with the Prime Mover/Dynamometer, Model 8960-1.

(Continued on next page)

Appendix C   Equipment Utilization Chart

Equipment		Exercise									
Model	Description	4-1	4-2	4-3	4-4	5-1	5-2	6-1	6-2	6-3	6-4
8134[1]	EMS Workstation	1	1	1	1	1	1	1	1	1	1
8211	DC Motor/Generator										
8221[2]	Four-Pole Squirrel-Cage Induction Motor	1	1	1	1						
8241	Three-Phase Synchronous Motor/Generator					1	1	1	1	1	1
8251	Capacitor-Start Motor				1						
8254	Universal Motor										
8311	Resistive Load		1			1		1	1	1	
8321	Inductive Load								1	1	
8331	Capacitive Load				1				1	1	
8621	Synchronizing Module										1
8821-2X	Power Supply	1	1	1	1	1	1	1	1	1	1
8942	Timing Belt	1	1	1		1	1	1	1	1	1
8951	Connection Leads and Accessories	1	1	1	1	1	1	1	1	1	1
8960-1 or -2	Prime Mover and Dynamometer Module	1	1	1		1	1	1	1	1	1
9061, 9062, or 9063	Data Acquisition Module	1	1	1	1	1	1	1	1	1	1

[1] Workstation model 8110-2 can also be used.

[2] Model 8221-2 can also be used.

Additional equipment

Completion of the exercises in this manual requires an IBM®-type 486 computer running under Windows®.

Appendix D

New Terms and Words

ac motor — An electric motor that operates from an ac power source.

armature — The rotating part of an electric motor or generator.

brake — Device or machine that opposes motion, either linear or rotational. The force opposing motion can be obtained through the interaction of magnetic fields in the brake that produces electrical energy. In this case, mechanical energy is converted into electrical energy by the brake, i.e., the brake operates as a generator. The force opposing motion can also be obtained using friction. In this case, mechanical energy is converted into heat by the brake.

brushes — Strips, blades, or blocks, usually made of metal or carbon, which are mounted on the stator of a rotating machine and provide sliding contact with the commutator or the slip rings of the rotor. Brushes allow current flow between the stator and rotor of a rotating machine.

commutator — Part of the rotor of a rotating machine (dc motor, dc generator, universal motor, etc.) that is made of many segments (parallel copper bars or strips insulated from each other) that are connected to the rotor windings. As the rotor turns, the segments successively make contact with brushes to distribute current to the rotor windings. The commutator converts dc current into ac current, or ac current into dc current, depending on whether the machine operates as a motor or generator

dc motor — An electric motor that operates from a dc power source.

dynamometer — A device which measures torque, as well as speed and mechanical power in certain cases, while motoring or braking the machine under test.

electric motor — A rotating machine that converts electrical energy into mechanical energy through the process of electromagnetic induction and interacting magnetic fields.

electromagnetic induction — The production of an electromotive force (emf), i.e., a voltage, in a circuit by a change in the magnetic flux linking with that circuit.

electromagnet — A device that produces a magnetic field when an electric current flows through it. A coil of wire wound around an iron core is an example of an electromagnet.

field current — A dc current which produces the fixed magnetic field in a rotating machine.

generator	A rotating machine that converts mechanical energy into electrical energy (either ac or dc) through the process of electromagnetic induction.
magnetic force	The force of attraction or repulsion between magnetic poles. Like magnetic poles repel each other, while unlike magnetic poles attract each other.
magnetic poles	The parts of a magnet where the magnetic lines of force exit, or enter, and where they are the most concentrated. By convention, magnetic lines of force exit from the north magnetic pole and enter at the south magnetic pole.
magnetic torque	The torque caused by magnetic forces.
motor efficiency	The ratio of the mechanical power (P_m) delivered by a motor to the electrical power (P_{IN}) supplied to the motor, P_m/P_{IN}.
motor power	The mechanical power (P_m) delivered by a motor, expressed in watts (W). Motor power is obtained by dividing the product of the motor speed n and torque T by 9.55 ($P_m = n \times T/9.55$) when the speed and torque are expressed in r/min and N·m, respectively. The product of speed and torque is divided by 84.51 when the torque is expressed in lbf·in.
output torque	Torque available at the shaft of a prime mover, such as an electric motor, to produce rotation.
prime mover	The primary source of mechanical power for any mechanical system that requires force to drive gears, belts, flywheels, etc.
rectifier	Electronic component that converts ac power into dc power.
rotor	The rotating part of an electric motor or generator.
slip	In rotating machinery, the difference between the speed of the rotating magnetic field and that of the rotor. Slip can be expressed in revolutions per minute (r/min) or as a percentage of the synchronous speed.
speed	The number of turns per unit of time at which a motor or generator rotates. Speed is usually expressed in revolutions per minute (r/min).
stator	The non-rotating part of an electric motor or generator.
torque	The twisting force applied to an object. Torque can be expressed in Newton·meters (N·m) or in pound force·inches (lbf·in). Electric power applied to a motor produces torque that makes the motor turn, and a generator turns because of the torque applied to its shaft by a drive motor, belt, or gear.

Index of New Terms

 The bold page number indicates the main entry. Refer to Appendix D for definitions of new terms.

ac motor **1**

brake **7**, 14, 31, 99, 197
brushes **12**, 30, 61, 68, 124

commutator **61**, 124

dc motor **1**, 11, 29, 45, 61, 67, 81, 99, 121, 123, 139

electric motor **1**, 64
electromagnet **2**, 7, 61, 67, 81, 100, 123, 139, 158, 159, 197, 213, 235, 257
electromagnetic induction **6**

field current **67**, 81, 100, 121, 125, 139, 215, 248, 266

generator **6**, 61, 99, 123, 173, 235, 237, 247, 257, 265

magnetic force **1**, 158, 185
magnetic poles **2**, 64, 102
magnetic torque **12**, 30, 124
motor efficiency **45**
motor power **45**

output torque **14**, 31

prime mover **11**, 29, 45

rectifier **65**
rotor **1**, 7, 45, 61, 67, 139, 158, 160, 185

speed **9**
stator **7**, 64, 67, 102, 139, 160, 185

torque 7, **8**, 31, 45, 121

Bibliography

Jackson, Herbert W. *Introduction to Electric Circuits*, 5th edition, New Jersey: Prentice Hall, 1981. ISBN 0-13-481432-0.

Wildi, Theodore. *Electrical Machines, Drives, and Power Systems*, 2nd edition, New Jersey: Prentice Hall, 1991. ISBN 0-13-251547-4.

Wildi, Théodore. *Électrotechnique*, 2e édition, Sainte-Foy: Les Presses de l'Université Laval, 1991, ISBN 2-7637-7248-X.